UNCHALLENGED PRIVILEGE

UNCHALLENGED PRIVILEGE

The Billion-Dollar Trilateral
Gravitational-Wave Discovery Scam

An investigative report by
Bibhas De
Author of *The Falsifiers of the Universe*

Printed in the United States of America
Books by Bibhas De
First edition published: August 2016
www.bibhasde.com/unchallengedprivilege.html

Copyright ©2016 by Bibhas R. De
All rights reserved.
ISBN-13: 978-1535349666
ISBN-10: 1535349662

The logos on the front cover have been reproduced under the fair use doctrine.

In a time of universal deceit, telling the truth is a revolutionary act.

George Orwell

CONTENTS

	PREFACE	i
1	Introduction: A quest for a waveform	1
2	The instrument: The proton yardstick	4
3	The discovery: That telltale waveform!	8
4	The discovery scam: High-tech hocus-pocus	11
5	The botched instrument: Left out in the rain	25
6	Dissidence: Slow thinking in fast times	32
7	Assignment of blame: *Enlightened people...*	38
8	LIGO India: A nation disserved	44
9	Concluding remarks	47
	NOTES AND REFERENCES	49
	ABOUT THE AUTHOR	53

PREFACE

This book is being published after nearly six months of my posting the scientific analysis now in the book on my Internet blog site *The Dreamheron Chronicles* (dreamheron.wordpress.com). During this period the LIGO discovery has been forcefully and urgently installed worldwide, with one confirming follow-up discovery and a barrage of scientific "idea" papers that continue to build on and bolster those discoveries. There has arisen a global stratagem to resurrect Big Bang Cosmology on LIGO's back – the cosmology that I thought I had laid to rest. Prize givers have been falling over each other to anoint LIGO. Taxpayer moneys have continued to flow apace to the LIGO cause both in the US and in India. These developments solidified my resolve to create a comprehensive physical document (and not just isolated electronic posts). Hence this book.

As was the case with my previous book *The Falsifiers of the Universe* (2015), this book too is an investigative report – to be distinguished from a scientific critique. The language of the book is direct and unsparing. As with my previous book I do not address esoteric mathematics or highfalutin theory. Instead I simply take my do-it-yourselfer's toolbox to the experimental contraption that installed the discovery.

The title of this book derives from the famous statement of Noam Chomsky: *Moral cowardice and intellectual corruption are the natural concomitants of unchallenged privilege.*

The part about intellectual corruption is obvious in the present context. But where is moral cowardice?

Those who seek scientific glory but do not want to try to get there the honest way, and instead form a powerful and well-backed collective within which they can scam together with impunity are, in my view, cowards of a kind. For true science requires true moral courage: courage to negotiate enervating failures, courage to own up to mistakes, courage to entertain self-doubt, and courage to pick up again and begin again. People in the latter category, the working stiffs of science, seldom achieve fame and fortune. People in the former category, the glory hounds of science, readily do. And when they are exposed beyond repair, the same groups that once backed them now actively cover up for them. The media which once promoted them to high heaven now simply goes mum. If this is not collective moral cowardice and societal rot, what is?

1 August 2016

1. Introduction: A quest for a waveform

Laser Interferometer Gravitational Wave Observatory (LIGO) has a long and storied history. It is today one of the most expensive and longest-running scientific research projects in the United States. It has been funded throughout its existence by the American taxpayers through the National Science Foundation (NSF). It is organized as a joint project mainly between the California Institute of Technology (Caltech) and the Massachusetts Institute of Technology (MIT). However, the LIGO Scientific Collaboration consisting of some 1000 members is made up of researchers from many American and international centers of research. The project is said to have cost thus far $1.1 billion, coming from NSF and international partners [1].

The most visible personalities behind LIGO in recent times have emerged to be Kip S. Thorne, Feynman Professor of Theoretical Physics (emeritus) at Caltech; Rainer Weiss, Professor of Physics (emeritus) at MIT; and the NSF Director France A. Córdova. The official operational head of LIGO is David Reitze, Professor of Physics at the University of Florida.

It is Thorne, a theoretician specializing on gravitation, who is credited with the inception of LIGO as early as 1968. Weiss, an instrumentation expert, later teamed up with Thorne. After various ups and downs, the project broke ground in 1994. Meanwhile a third member Ronald Drever, a Scotsman from Glasgow, had joined Thorne and Weiss. He joined the Caltech physics faculty. The trio would come to be known as the founders of LIGO. Other background information about LIGO will be presented as needed and where needed.

The plan of the book is as follows: The first three chapters will present faithfully the LIGO science as described by the LIGO scientists. This description will be kept limited to the bare essentials needed for the present discussion. There are two reasons for this economy. First, this keeps the analysis clean and simple and accessible. Second, this avoids burying fundamental issues under mountains of high-tech complexities. The investigative report starts with Chapter 4.

From time to time I will mention the Falsifiers book. This refers to my 2015 book *The Falsifiers of the Universe* [2]. This book covers a series of instrumentation frauds that sequentially installed Big Bang Cosmology as the official history of the universe. The COBE Satellite fraud is covered here in detail. Additionally, the book covers the BICEP2 instrumental botch up. This was a project aimed at finding signatures of Big Bang inflation era in the polarization of cosmic microwave background radiation, taken to be the Big Bang relic radiation that has a blackbody spectrum.

The purpose of the LIGO instrument is to try to detect gravitational waves from catastrophic and transient cosmological events that suddenly release enormous amounts of energy in the form of such waves. Gravitational waves are propagating compressions and expansions of space. A rough analogy would be sound waves which are propagating compressions and expansions of air. Space however is empty. But even it can undergo compressions and expansions. Gravitational waves are assumed to travel at the speed c (= 3.10^8 m/s), the speed of light.

There are a few conceivable strong sources of gravitational wave in the universe: supernova explosions, mergers of binary neutron stars and binary black holes, and the original Big Bang explosion itself. When these events occur, a portion of the constituent mass is converted to energy in the form of gravitational wave which spreads out in the universe. What we expect to observe on Earth is a transient wave of short duration, perhaps showing only a few cycles of the wave. Such a trace is called a waveform.

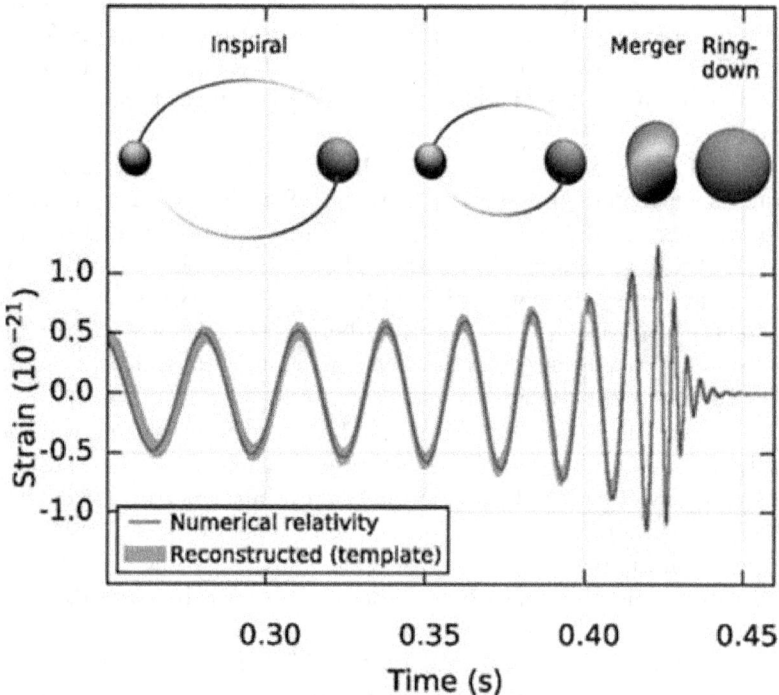

Figure I-1: Binary black hole merger and definition of the terms inspiral, chirp and ringdown. Chirp refers to the increase in frequency and amplitude with time.
[Image courtesy: Quora]

The LIGO researchers have developed theoretical models and computer simulations to predict what a gravitational waveform would look like from such an event. Here we are concerned only with the merger of two black holes. In this model two black holes of masses m_1 and m_2 approach each other as they spiral in [3]. Even as they do so, they begin to release energy in the form of the said wave. The frequency of the wave is twice the orbital frequency of the black holes. The shape of the wave, or the waveform, produced as the two black holes spiral and approach each other is labelled *inspiral* (Figure I-1). The closer the two bodies get, the faster is the orbital frequency and greater the energy released. Hence the frequency and the amplitude of the gravitational wave at this stage increase with time. This phenomenon is called *chirp*. When the black holes are near merger, they are moving at velocities close to the speed of light. Eventually, the two black holes physically merge into one, where the merged black hole ends with a mass somewhat less than $m_1 + m_2$. The difference in mass is what has been converted to energy in accordance with the mass-energy relation.

The merged black hole at this time has a highly deformed shape. As it assumes a spherical shape it continues to emit the wave with diminishing energy. This is the *ringdown* stage. After that everything settles.

Inspiral, chirp and ringdown — these are the three signatures of gravitational wave in an observed waveform. This combination cannot be produced by any other conceivable sources of the signal.

So this is the type of waveform the LIGO researchers would be searching in the data traces they gather as LIGO looks to the sky. This is their quest.

> The vision, persistence, and leadership of Ronald Drever, Kip Thorne, and Rainer Weiss, along with the contributions of a thousand collaborators on the LIGO discovery team, led to the first detection of gravitational waves, not only validating a key prediction of Einstein's general theory of relativity but inaugurating a new method for studying cosmology, in particular the workings of astronomical objects exhibiting the greatest gravitational effects in the universe.
>
> *from the statement accompanying the Gruber Cosmology Prize to LIGO*

2. The instrument: The proton yardstick

The object of the LIGO instrument is to detect and measure the compression and expansion of space as a gravitational wave train travels past the Earth. Consider the plane perpendicular to the direction of wave propagation. As the wave compresses space in one direction in this plane, it causes the space to expand in the orthogonal direction in the plane. The instrument is designed to detect this directional difference. It does so by employing the principle of laser interferometry.

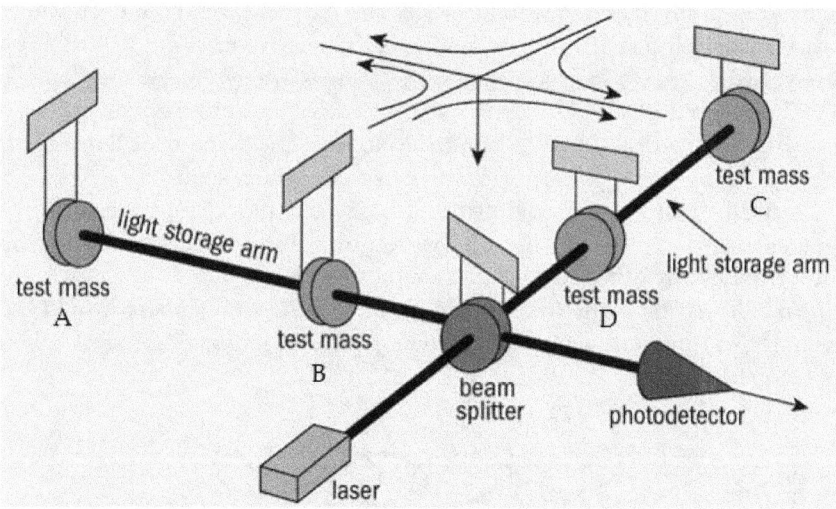

Figure II-1: Simplified schematic of the LIGO instrument: The length of the arm AB (=the length of the arm CD) = 4 km. The mirrors are made of fused silica with dielectric coating. Diameter of mirror A or C = 34 cm; thickness of mirror = 20 cm; mass of mirror A or C (test mass) = 40 kg; laser beam radius = 5.3 – 6.2 cm; wavelength of laser light, λ_{laser} = 1064 nm (near infrared); laser polarization = linear horizontal, laser power transmitted = 125 watts [*Image source: LIGO*]

Refer to Figure II-1. There are two orthogonal directions AB and CD. The distances AB and CD are exactly the same (= l). The test masses A and C are fully reflecting mirrors at the end of these two arms. The test masses B and D are partially reflecting and partially transmitting mirrors. There are a source of the laser beam, and a photodetector which detects the laser light incident on it.

Now we will follow the laser beam along its path as it leaves the source. As it hits the beam splitter, by the property of this device, half the power in the beam proceeds along the path BA and the other half proceeds along DC.

The first part of the light is reflected from A and hits the beam splitter. Half of this light now proceeds towards the source and half towards the photodetector.

The second part of the light is reflected from D and hits the beam splitter. Half of this light proceeds towards the source and half towards the photodetector.

The two parts of the beam that arrive at the source have traveled exactly the same distances in exactly the same time and are now exactly in phase. Therefore, they add to one another. In other words, they interfere constructively.

The two parts of the beam that arrive at the photodetector have traveled exactly the same distances in exactly the same time. But one beam has suffered an extra reflection and hence the two beams are now exactly 180 degrees out of phase. Therefore, they cancel one another. In other words, they interfere destructively. The photodetector sees zero light. It is dark.

So, because the orthogonal arms are of the same length, the light incident on the photodetector is zero. If the two arms become of unequal length for some reason, one beam will take slightly longer time (longer by Δt, say) than the other to arrive at the photodetector. It will now see a small amount of light because the phases of the arriving waves will differ from 180 degrees. There will be a nonzero phase shift $\Delta\varphi$ between the beams.

This is the rudimentary principle of the instrument. If a gravitational wave travels past it in an arbitrary direction, it will cause the length of one arm to be compressed (e.g. the mirrors A and B will be closer together) and that of the other arm to be expanded (mirrors C and D will be farther apart) at the same time. The resulting path difference Δl (= c Δt) between AB and CD will cause an amount of light to shine on the photodetector. This bit of light is used by the instrument to quantify the movement Δl. This quantity is related to the phase shift by $\Delta l = (\Delta\varphi/4\pi) \lambda_{laser}$.

All this is what happens at one instant of time. At the next instant Δl will change as the wave moves a tiny bit further past the detector. Now Δl will be determined again. In this way Δl (plotted against the time t) will trace out the passage of the wave, essentially mimicking the wave.

The longer the arms are, the larger is Δl and hence the larger the phase shift and the larger amount of light that arises at the photodetector. More light here means better and more accurate measurement and better threshold of detection. This is the incentive for making the arms AB and CD as long as practical.

The arms of LIGO are encased in 1.2 m diameter steel tubing sealed at the ends. The space inside the tubing is evacuated to a high degree of vacuum in order to prevent gas molecules from degrading the laser beam and causing unwanted vibrations of the mirrors.

There are here two types of movement we are concerned with:

(1). The mirror moves because of natural (e.g., seismic) or man-made vibrations, or because of the effect of currents generated in the metal structures due to geomagnetic disturbances.

(2). The mirror "moves" because of the passage of a gravitational wave. This is not a conventional movement. It is the space itself that is expanding and contracting along with everything that is in it. So the mirrors A and B for example will be closer together sometime and farther apart sometime compared to where they were before the arrival of the wave. The same is true of the mirrors C and D.

LIGO presents its measurements in terms of *strain* which is the fractional change in length of a LIGO arm (strain $h = \Delta l / l$). A typical black hole merger event has been estimated theoretically to produce a strain of 1 part in 10^{21}. This degree of smallness of the signal has been communicated to the general public by the LIGO scientists as follows: The LIGO instrument needs to be able to measure a mirror movement that is less than 1/10,000-th the diameter of a proton. While the diameter of a proton is a complex concept, for our purposes here is about 1.6×10^{-15} m. It is important to bear in mind that we are speaking here of macroscopic movement, and not movement in subnuclear quantum physics, for example. So this ability of LIGO to function as a "proton yardstick" is truly an astounding feat. LIGO has in fact been described as the most precision scientific measuring instrument ever built.

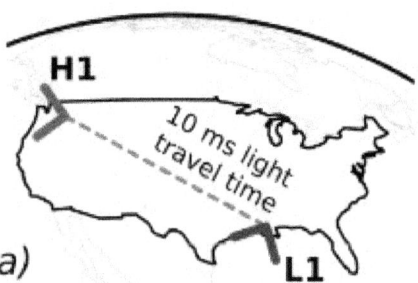

Figure II-2: Location and orientation of the Hanford and Livingston detectors. Notice that determining the relative three-dimensional orientation of the two detectors must include the curvature of the Earth.

The instrument as described cannot however measure a strain this small. To remedy this situation, the laser beam is made to bounce back and forth between two mirrors (A and B; and C and D) some 300 times before arriving at the photodetector. This effectively enlarges the 4 km arm to 1200 km. The bouncing laser beam accumulates the displacement Δl three hundred times, bringing this amplified displacement within the range of measurability.

This is the basic description of LIGO. In practice the instrument is far more complex. These complexities of the instrument give rise to much higher precision and reliability with which the measurements can be made. But those complexities do not enter our discussion. All we need to carry in mind is the basic technique by which LIGO measures the movement of the mirrors.

The discovery of gravitational waves involved two LIGO stations that are identical (instrument-wise). One is located in Hanford in the state of Washington, and the other in Livingston in Louisiana. The idea of having two stations 3000 km apart is to test whether a signal observed is recorded simultaneously at both stations. A gravitational wave will produce the same signal (same waveform) at the two stations. Other types of disturbances generally will not.

The light travel time for the 3000 km path – the straight-line distance between the two stations - is 10 ms. So if a gravitational wave arrives in a direction parallel to the line joining the two stations, it will arrive at the second station 10 ms after it arrives at the first station. If the wave is traveling perpendicularly to the line joining the two stations, the delay time will be zero. If the wave arrives at other angles, the delay time will be between 0 ms and 10 ms. The actual delay time then can be used to estimate the angle of arrival, identifying a circle in the sky on which the source of the wave is located. With additional observational data the location can be narrowed down. If there were a third station, the location can be pin-pointed on the sky.

> The tiny effect that gravitational waves have on space led many scientists to believe they would be undetectable. A breakthrough was achieved in 1972, when Weiss worked out the basic interferometer concept that eventually became LIGO. Weiss provided technical leadership and devoted his extraordinary experimental acumen over the next decades, contributing to every aspect of the final apparatus.
>
> *from the statement accompanying the Kavli Prize in Astrophysics for LIGO*

3. The discovery: That telltale waveform!

On 14 September 2015 at 5:51 a.m. Eastern Daylight Time (09:51 UTC) there were recorded almost simultaneously at the Hanford and the Livingston detectors two waveforms that looked nearly identical. They were displaced in time by about 7 ms, the Livingston signal arriving that much later. This observation was reported on 11 February 2016.

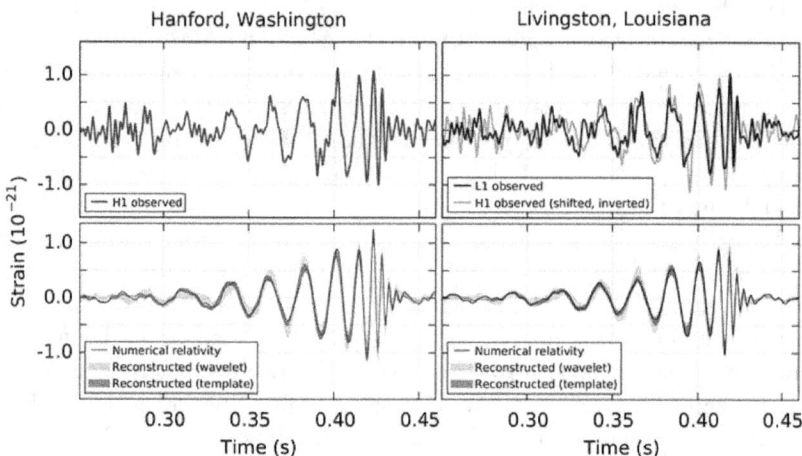

Figure III-1: The first detection of black holes merger. Comparison of the amplitudes observed by two identical stations at Hanford WA and Livingston LA.
Top panel: Observational data for Hanford (left); and Livingston (right, dark line) compared with Hanford (grey line, inverted from the diagram on left for proper comparison, and time-shifted by 7 ms).
Bottom panel: Theoretical prediction for binary black hole merger. The Livingston signal is weaker because of different orientation of the detectors, as shown in Figure II-2. [*Image source: LIGO*]

Refer to Figure III-1. A number of things about this twin observation suggested a gravitational wave to the LIGO scientists. First, the nearly identical form of the two signals suggested the passage of the same disturbance first through Hanford and later through Livingston. They assumed gravitational waves travel at the speed of light, and have a planar wavefront. They then calculated that the 7 ms time lag corresponds to the arrival of the wave at a certain inclined angle relative to the line joining the two stations. From this angle they identified a broad patch of sky from which the wave likely originated.

Next they studied the details of the wiggles and compared the wiggles with the library of templates they have for the expected nature of the gravitational waveform from various sources. There was close resemblance to the scenario where two spiraling black holes were approaching each other and eventually collapsing into a single black hole.

They made theoretical calculations and found that the scenario that best matched the observed waveforms were two black holes with masses equal to 36 and 29 solar masses located about 1.3 billion light years away, spiraling and merging together to form a single black hole of 62 solar masses (cf. Figure I-1). About 3 solar masses transformed to energy which propagated away as gravitational wave. The observed waveform thus showed the inspiral stage, the chirp, and the ringdown of the coalesced black hole. The chirp frequency rose from 35 Hz to 250 Hz.

With all these corroborations, it was concluded with a level of confidence approaching certainty that a gravitational wave and a black hole merger event had been detected.

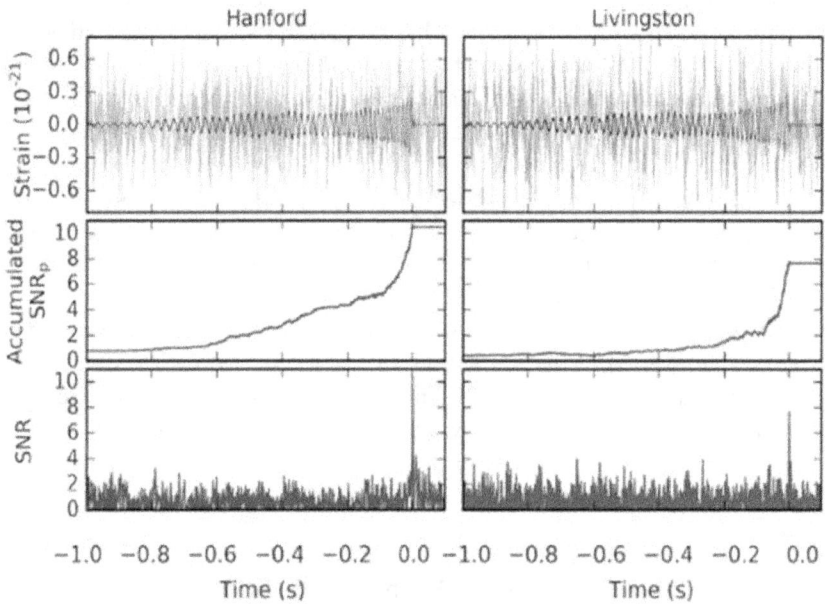

Figure III-2: The second LIGO detection of black hole merger. *Top panel*: Actual observational data (hash) and the best-fit merger scenario for the Hanford and Livingston stations. *Bottom panel*: Signal-to-noise ratio. *Middle panel*: Accumulated signal-to-noise ratio.

On 15 June 2016 a second discovery of black hole merger was reported [4]. It was an event recorded on 26 December 2016 at 3:38 a.m. UTC (9:38 p.m. CST on Christmas Day). This discovery took place along the lines of the first discovery, but with important differences. The signal this time was much weaker and did not stand out above the noise threshold. It had to be teased out of noise with advanced digital signal search and signal processing techniques. This time the merger involved black holes of 14 and 8 solar masses and the final merged black hole had 21 solar masses. One solar mass was converted to gravitational wave. The signal in this case arrived first at the Livingston site and 1.1 ms later at the Hanford site. The chirp in this case was much less pronounced. Whereas in the first discovery only a few inspiral cycles were seen in the diagram, in the present case some 55 cycles were seen – corresponding to half as many binary orbits.

This is a very simplified account of the two discoveries, but that is all we will need.

The presentation of the LIGO discoveries to the world at large as well as to the expert scientific establishment was simply superb. No one could have asked for greater perfection. There was intricate theory. There was hairy mathematics. There was esoteric physics. There was the most advanced computer power. All these were folded together to provide detailed models of events that occurred more than a billion years ago and greater than a billion light years away. The models were presented as understandable xy graphs with a number of discriminating features. And then came in cutting-edge technology to execute the idea. When the discoveries were unrolled, everyone saw those very xy graphs with those very discerning features – only this time not from a computer modeling a theory, but straight from the LIGO observatory looking skyward.

But even so, there was raised some doubt in some quarters after the first discovery report. After the second discovery, it was said by LIGO members and promulgated in the media that this discovery silenced all the sceptics who had said the first discovery was a fluke. The second discovery was said to have removed any and all scientific doubt about LIGO. It was now on to the era of full-blown gravitational wave astronomy.

> ### RAINER WEISS AND HIS MONKEY: A *GREEN TEA* STORY!
>
> I feel an enormous sense of relief and some joy, but mostly relief. There's a monkey that's been sitting on my shoulder for 40 years, and he's been nattering in my ear and saying, "Ehhh, how do you know this is really going to work? You've gotten a whole bunch of people involved. Suppose it never works right?" And suddenly, he's jumped off. It's a huge relief.
>
> In another interview:
>
> I feel like a monkey just jumped off my back! But the monkey's not gone yet, he's still walking along here on the sidewalk.
>
> *Rainer Weiss on the LIGO discovery*

4. The discovery scam: High-tech hocus-pocus

The LIGO instrument and its functioning are flawed in many ways and in many layers. One fault can make discussing another fault unnecessary or redundant or no longer relevant as far as discarding the discovery is concerned. But for the sake of method and order I will discuss everything.

A class of experimental scientific instruments may be symbolically or schematically described as a two-port device, with an input port, an output port and a connection for providing power etc. to the device. The word port here is symbolic, but it may be an actual physical port (e.g., a radiofrequency connector). We consider the specific case where the device is receiving a waveform or a pulse – a transient signal whose amplitude varies with time.

First we want to discuss a conventional example to set the stage for discussing the LIGO instrument, since the latter instrument is largely in an unknown and uncharted territory of scientific experimentation.

Figure IV-1: A symbolic two-port scientific instrument.

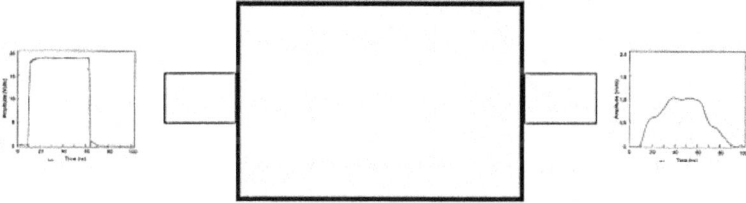

Figure IV-2: The input waveform (an electromagnetic pulse) and the corresponding output waveform for my experiment. The waveforms are presented in an amplitude-vs-time diagram. (See Figure IV-3 for an enlarged view.)

For the sake of specificity of discussion, I have chosen an illustrative example from a past study of my own [5]. I will describe the actual instrument later. For this experiment, the input and the output are two electromagnetic pulses (waveforms) as shown in Figure IV-2. We can see immediately that the two are drastically different. Generally speaking, the input and the output of any such instrument are never the same before any remedies have been applied. Only the degree to which they differ vary.

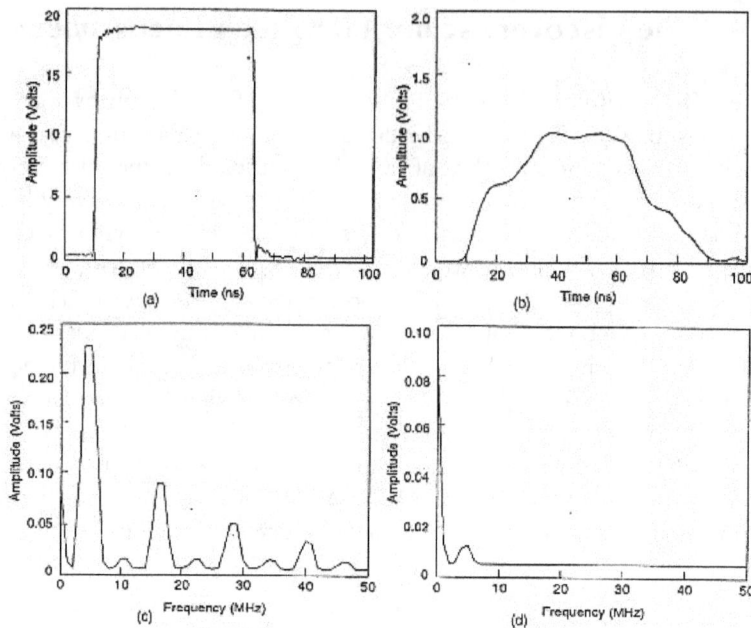

Figure IV-3: *Top panel*: The input and the output pulses for Figure IV-2. *Bottom panel*: The frequency content of the pulses.

Figure IV-3 shows the frequency content of the input and the output waveforms. We can see that the effect of the instrument here is to selectively take out the higher frequency content. This effect of the instrument – described by an instrument transfer function – can be understood as follows.

Let I(t) be the input pulse, taken in its entirety (starting from zero baseline and coming back to zero baseline.)

Let T(f) be the instrument transfer function (f = frequency).

The signal I(t) can be decomposed into its constituent frequencies:

I(f) = ∫I(t) exp (-2i π ft) dt.

At the output of the instrument I(f) is modified to

O(f) = I(f)xT(f).

Now the output signal is

O(t) = ∫O(f) exp (2i π ft) df.

This is how I(t) is transformed to O(t) by the instrument. Thus the transfer function – a property of the instrument which can be determined from a control experiment with known input and output waveforms – gives us a way to subsequently determine an unknown output waveform for a known input waveform or vice versa.

In all such experiments where one waveform is known and the other waveform is to be determined, the transfer function must be determined and reported. This scientific necessity cannot be replaced by declarative statements such as There is no loss, There is no distortion, There is no noise... etc. Nor can its role be replaced by doing various digital processing of the output signal. The transfer function must be determined and reported in some manner if the study is to have any scientific validity. In the above experiment, the bottom panel of Figure IV-3 constitutes that report. The instrument transfer function (in this case numerical) is readily calculated from the frequency-domain transforms.

Now, for the purpose of developing the above point, let us make up an ad hoc experimental scenario, with a cartoon even. Refer to Figure IV-4. Suppose that a certain type of cosmological source has been theorized and it has been predicted to emit a stream of electromagnetic waveforms in the shape of a skewed Gaussian pulse. Suppose a radio telescope is scanning the sky in the survey mode and the output of the antenna is being brought down to the observatory housing for detection and analysis. Suppose the housing is about 1 km away. Whatever is involved in transferring the signal from the radio telescope terminal at the focus of the parabola to the detector in the house is inside the box defined in Figure IV-2.

Figure IV-4:
An ad hoc experimental scenario. A radio telescope performs a sky survey looking for a new type of cosmological source that puts out pulses with a skewed Gaussian shape. The input to the box is the signal captured by the telescope. The output of the box is the detected pulse in the observatory housing.

One day a skewed Gaussian is indeed seen. The free parameters in the theory are tweaked to provide an impressive fit between theory and observation. A grand discovery results, with a phenomenal level of confidence..

What is wrong with this great discovery? This proof assumes that the unknown waveform received by the radio telescope and appearing at its terminal is the same as the output waveform observed in the housing. But the black box containing everything that brings the waveform to the control panel has an instrument transfer function. Had the transfer function been applied to the observed waveform to determine the actual incident waveform, there would be no discovery. Instead we would derive a boxcar-shaped waveform at the telescope terminal – a waveform unlikely to be produced by any natural sources.

Before I go on to the LIGO experiment, I will describe what was in the box in my original experiment. I needed to find out what happens to an electromagnetic pulse after it is transmitted through a long run of fiber-optic cable. So, the box essentially has a radiofrequency coaxial connector as the input port, and another such connector as the output port. Following the signal along its path, the first component inside the box is an opto-electronic device that transforms the electromagnetic pulse to a modulation of the light wave traveling along the fiber. After a 0.9 km run of the fiber, there is another opto-electronic device that reconstitutes the pulse and delivers it at the output port as an electromagnetic pulse.

There is no suggestion here that the equipment inside the box in this experiment has any analogy to LIGO. That is precisely the point. There could be anything inside the box – electrical, electromagnetic, optical, electro-mechanical, opto-electronic, ... components - and also feedback loops, servo motors and even software. As long as there is an input and an output and the device is working, there is an instrument transfer function.

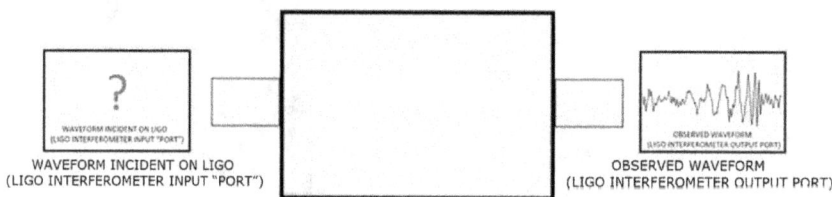

Figure IV-5: The output waveform detected by the LIGO instrument. The input waveform is unknown. To find it, we have to work back from the output waveform, using the instrument transfer function. But that function for LIGO is indeterminate.

Now let us establish the analogy of LIGO to the two-port instrument. LIGO does not of course have a physical input port. The gravitational wave is incident on its distended arms over a large expanse of land. The LIGO waveform amplitude is in strain rather than volts as in the above case. And the LIGO pulse is transient rather than repetitive. Even with all these differences, schematically we can represent LIGO as a two-port instrument.

The fact that there are two arms of LIGO is not to be seen as it having two input ports. There is one input port and there is one output port. Everything else is inside the box.

With the LIGO instrument in the discovery mode one is obtaining the output waveform. The input waveform is the unknown thing to be figured out as the first step of the discovery process. To find what it is, one needs to have performed earlier controlled experiments where both the waveforms are known. From this study, an instrument transfer function would have been calculated and applied to the discovery mode operation.

However, this controlled experiment cannot be performed because the conditions of a gravitational wave incident on LIGO cannot be experimentally realized. How would one, for example, make a LIGO mirror move by 1/10,000-th the diameter of a proton in a controlled experiment in the manner a gravitational wave would move it?

Thus the transfer function is indeterminate. Therefore LIGO is not a viable discovery instrument.

But it seems that the LIGO team proceeded regardless with what appears to be an assumption that $T(f) \equiv 1$ for all values of f in the relevant frequency range. This leads to the conclusion that the input and the output waveforms are identical. The measured output waveform can then be compared directly with the theoretical prediction of the gravitational waveform incident on LIGO. This is indeed what was done and what made the discovery possible.

There has been submitted by the LIGO team no evidence or proof that $T(f) \equiv 1$ over all frequencies of interest. They have done some laborious "calibration" experiments to this end [6, 7]. But these experiments have no bearing on the LIGO instrument transfer function. For example, they tried to simulate the action of a gravitational wave on a LIGO end mirror by shining an auxiliary laser beam on the mirror (Figure IV-6). The light (photon) pressure then causes the mirror to move and the interferometer detects this movement. The auxiliary laser beam is modulated to provide a waveform-like calibration signal.

This experiment has little relevance to the case of LIGO observing a gravitational wave. In this experiment the mirror moves according to the ordinary laws of motion which include the mirror mass. In the gravitational wave case, the mirror "moves" differently. The waveform shapes from the two causes would consequently be different. As we have seen, the waveform shape is of paramount importance in the LIGO discovery. Furthermore, in the calibration experiment the result does not necessarily depend on the length of the arm whereas in the discovery experiment it does. Also, only one mirror is actuated in this experiment whereas the LIGO principle rests fundamentally on concerted dual-mirror actuation. There exists no experimental support that such actuation actually occurs.

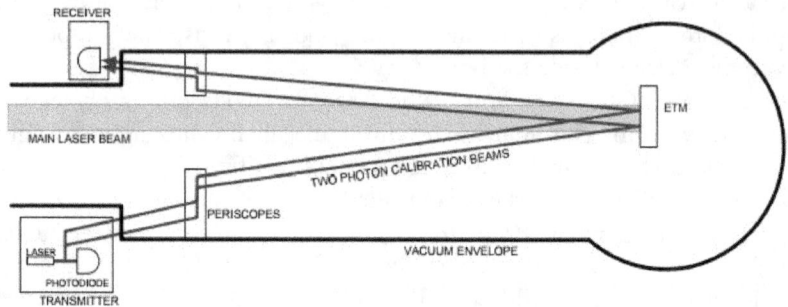

Figure IV-6: A calibration method for LIGO. An auxiliary laser beam is used to push on the mirror (End Test Mass or ETM) to cause it to move by minuscule amounts. The recoil of the light off the mirror causes the mirror to be displaced. The original beam from the auxiliary source is spilt into two equal power beams that hit the mirror above and below the patch illuminated by the main laser beam. This configuration avoids interfering with the main beam by elastic deformation of the said patch, and also avoids tilting the mirror. The LIGO instrument senses and records this displacement the same way it senses and records the displacement caused by a gravitational wave. The auxiliary laser wavelength is 1047 nm and puts out continuous-wave power of 2 Watts. This power can be sinusoidally modulated by as much as 1 Watt peak-to-peak, thus creating a waveform-like signal. The displacement caused is ~ 10^{-16} m-rms at 100 Hz [6].

    This is a good time to pause to take in an important point about the LIGO instrument: The actuation of LIGO may occur in a continuum of states. At one extreme is the state where only one end mirror is actuated and the other end mirror remains stationary (as is the case with the above calibration experiment). At the other extreme is the state where both the end mirrors are actuated in a coordinated way (as is the case with gravitational waves). In between these two extremes are states where the end mirrors are actuated to varying degrees, not necessarily in a coordinated way. Such states will arise for example from geomagnetic disturbances.

    In all these cases LIGO will report an output waveform. But one cannot say by looking at this waveform what the actuation state was. So to identify an output waveform as being due to gravitational wave requires convincing independent evidence that both end mirrors were actuated. Without this piece of the overall proof, the discovery remains incomplete. This is a pure instrument issue. However, there are so many issues with LIGO design that it is pointless to try to fix any single one of them.

This is also a good place to examine the central assertion that LIGO can measure displacements of the end mirrors caused by the passage of a gravitational wave down to ~ 1/10,000-th the proton diameter. So this displacement is $\Delta l \sim 10^{-19}$ m. In the experiment in Figure IV-6, the displacement achieved is $\Delta l \sim 10^{-16}$ m. So the assertion and the achievement are nowhere in the same displacement regime. The experiment does not provide any support that the instrument functions way down in the length scale at the level of the proton yardstick. The LIGO instrument was uncalibrated when the two discoveries were made.

How do they determine Δl? It is found from the information the interferometer provides as we discussed in Chapter 2. This quantity is so small that it is below the threshold of measurability for LIGO. So they accumulate this small displacement by probing it repeatedly. The laser beam bounces off the end mirror N times (N ~ hundreds) to accumulate the small displacement of this mirror. So what the LIGO instrument actually given to measure $N\Delta l$ - a quantity that is reportedly above the measurement threshold of the system.

- A GW with $f_g \sim 100$ Hz $\Rightarrow \lambda_g \sim 3000$ km produces a tiny strain $h = \Delta L / L$
- We measure $\Delta\phi = 4\pi \Delta L/\lambda_{laser} = 4\pi L h /\lambda_{laser}$ so to measure small h, need large L
- But not too large! If $L > \lambda_g/4$, GW changes sign while laser light is still in arms, cancelling effect on $\Delta\phi$
- Optimal: $L > \lambda_g/4 \sim 750$ km. But not very practical!
- For more practical length (L ~ 4 km), increase phase sensitivity: $\Delta\phi = 4\pi \Delta L/\lambda_{laser} \Rightarrow \Delta\phi = N(4\pi \Delta L/\lambda_{laser})$, with N ~ 200
- N : Increase number of times light beam hits mirror, so that the light is phase-shifted N times the single-pass length diff ΔL

LIGO-G000165-00-R   AJW, Caltech, LIGO Project

Table IV-1: A LIGO slide showing the reasoning on how many laser bounces to use for optimum amplification of Δl. [*Source: LIGO*]

Table IV-1 [8] shows LIGO criterion on how to determine the optimum value of N at a design frequency for gravitational wave of 100 Hz (wavelength 3000 km). In this case N ~ 200. In actual LIGO discovery situation, the value of N used was 300. Therefore the design frequency of the gravitational wave according to the above criterion was ~ 67 Hz. So LIGO was functioning reliably in this respect for frequencies up to 67 Hz, and not reliably above this frequency. Therefore, there was no basis to deduce a chirp of 35 -250 Hz from the observed waveform. Therefore all other conclusions were invalid.

Thus, based solely on the criterion of Table IV-1 the LIGO discovery never happened.

However, this is not the only applicable criterion. Refer to Figure III-1, top panel. The observed waveforms here are shown as continuous lines, but data-wise they are dotted lines, presumably with a high density of dots to elicit the detailed undulations the discovery makes use of. Each dot represents one measurement data point, and corresponds to an instantaneous position of the mirror. Initially, we assume these waveforms to be faithful experimental representations of the incident signals and examine them from this point of view.

According to the LIGO operating principle, a complete 300-bounce measurement takes place at each dot. So the requirement here is that while this 300-bounce measurement takes place, the moving end mirror can be considered substantially stationary at its instantaneous position. Let T be the approximate time from a trough to the next peak of the observed waveform, and let τ be the time it takes to complete the 300 bounces. Then the above requirement translates to T » τ. In other words, between the trough and the peak, there will be T/τ dots to define the waveform.

For the LIGO waveform near the peak signal, T ~ 5 ms (Figure III-1, top panel; 0.420-0.425 s, for example). The quantity τ is simply the back-and-forth light travel time along an arm (4 km long) 600 times. So τ ~ 8 ms.

Thus in actual LIGO operation, T ~ τ. While the researchers think they are accumulating the displacement of the mirror at a fixed position (i.e. acquiring a specific measurement dot), the mirror has completed a large part of its movement cycle.

The waveforms reported by LIGO cannot resemble anything incident on it. The output is a highly corrupted version of the input.

Let us return to the calibration experiment. In spite of this experiment not being representative of the gravitational wave situation, if we still want to know what the above experiment found, we would learn that:

The calibration uncertainty is less than 10% in magnitude and 10^0 in phase across the relevant frequency band 20 Hz to 1 kHz.

So when on 14 September 2015 LIGO reportedly observed a gravitational wave, the instrument stood backed by an up-to-date record book of calibration results indicating uncertainties of 10% in magnitude and 10^0 in phase.

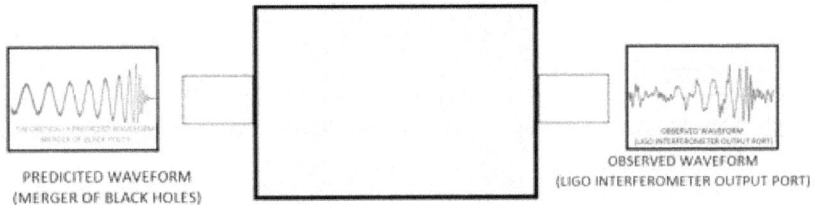

Figure IV-7: The unknown input waveform in the LIGO instrument is deep sixed and unjustifiably replaced by the theoretically predicted waveform. The grand discovery follows.

The reader with some background in waveform analysis, detection etc. might put this question to his gut: What can cause the smooth undulations (Figure IV-7) of the presumed input waveform to acquire the sharp and the rugged character of the observed output waveform? Do the uncertainty values quoted reconcile the two waveforms? No. Can incoherent noise from various sources explain this modification? No. The two waveforms simply do not belong together in the LIGO context as an input-output pair, their tantalizing similarity notwithstanding. And what can cause a heavy vibration-isolated pendulum to execute such jagged motion?

Next refer back to Figure III-1. Here we see that the peak amplitudes observed at the two stations differ by as much as 70% - far in excess of the instrumental uncertainty of 10%. The signal is definitely weaker in Livingston than in Hanford – very significantly so. But there is no mechanism by which a gravitational wave would be attenuated in traveling from Hanford WA 99343 to Livingston LA 70754.

The LIGO scientists have explained the amplitude difference between the two stations as the result of the different orientation of the two detectors (Figure II-2). First the incoming angle for the gravitational wave was calculated from a time delay of ~ 7 ms between the two stations. Then the amplitude difference was calculated from the orientations.

But as we see clearly in the Figure III-2 by comparing the top and the bottom panels, this theoretical orientation effect (bottom panel) does not explain the extent of the observed difference between the two signals (top panel). Furthermore, by concentrating only on the observed data, top panel right, we see that the difference between the two curves cannot be due to a cosine-like geometric projection effect. If this were the operative effect, there would be clearance between the curves all along, not just at the peaks.

Actually these LIGO orientation effect calculations are not scientifically valid. The orientation effect depends entirely on what the angular pattern of the LIGO antenna is for gravitational wave. This is a concept similar to that in electromagnetics or acoustics.

LIGO has a directional pattern that depends on the instrument design. There are many reasons why this pattern is not an ideal cosine-type pattern assumed above. Generally speaking, when an antenna is much smaller than the wavelength, its pattern tends towards being isotropic. LIGO gravitational wave wavelength for the first discovery is ~ thousands of kms, and LIGO size is ~ kms.

Therefore, in absence of information showing what the pattern is, orientation cannot be used to explain the difference in amplitude. And this pattern for LIGO is indeterminate. This is one more reason why LIGO is not a viable instrument of discovery.

The amplitude difference is a conundrum, and it arises because a gravitational wave is assumed to have been observed. Otherwise there is no issue.

The missing waveform is the missing link. The theory waveform should have been compared with the missing waveform. By deep sixing the role of the latter, the discoverers severed the scientific link between theory and experiment. Thus freed them from inconvenient scientific constraints, they were then able to game the scientific system by playing on the superficial resemblance of their theory waveform and the observed waveform. This resemblance is partly coincidental and partly concocted. Remember that the black hole merger calculation has great tweaking leeway. The mass of each black hole, the distance to the binary black hole etc. are all free parameters.

Now we move on to the second LIGO discovery. Here, perhaps in a hasty attempt to follow the first discovery with a confirmation, the faults with that discovery were compounded by introducing new faults. A presumptive template was applied to search for gravitational signal amid noise. A long train of waveform buried in the noise was identified in both stations that matched such a template. Different filters were applied to the data to correct for different instrument effects. Then the successful template itself was presented as the discovery. The actual observed waveforms - the counterpart of the top panel of Figure III-2 - never emerged from this study. We linked the observed noise hash to a neat theory template through some digital processing.

So we do not know if, for example, the chirp was present in those absent "counterpart" waveforms or if it was imposed on the measurement data by the wishful template. This procedure would not have passed as a scientific discovery had it not been presented amid the atmospherics and the hysterics created by the first discovery. Stagecraft, not science, was the key.

To me this was a "déjà vu all over again" situation. When there is fatal fault with the instrument on the ground, how can one repeatedly roll out phenomenal discoveries in the sky and aver that everything is fine? The BICEP2 scientists were doing exactly the same thing. Here is what I wrote about that in the Falsifiers book:

The BICEP2 team ushered in a new era of telescope science. They kept presenting assorted polarization maps and averring that the images are "on the sky", "on the sky" ... (meaning they are true sky features, and not instrumental artifacts.) When erroneous physics and engineering are plainly evident in the instrument on ground, how one could point to sky images to assert that everything is fine is something I never learned during my stint in radio astronomy or satellite communication.

This comment applies in toto to LIGO.

The second set of LIGO wiggles – if they are meaningful – arose for the same reason the first set of wiggles arose. Neither set has anything to do with any gravitational waves.

Finally, take another look at the calibration experiment. Although this did not reproduce the condition of a gravitational wave, it came closest to testing the overall performance of LIGO. There was a known input strain waveform, and there was a known corresponding output waveform. So the scientifically most essential and visually most convincing thing to do was to present these two waveforms either overlaid or over-under or side by side. If LIGO were functioning the way they say it was functioning when the discovery was made, these two waveforms from the controlled measurement would match perfectly. So why did they not show this comparison which would have put all instrumental criticism to rest?

The reason may be that the waveforms did not match at all, proving LIGO to be a botch up. Actually the other information they present from that experiment suggests that the two waveforms were not alike. So they may have suppressed "inculpatory evidence".

The last point of this chapter is that two-station simultaneous observation of an unknown signal does not strongly identify it to be gravitational wave. Other long-range effects are conceivable. In fact the two-station coincidence is the criterion Joseph Weber used to report his discovery of gravitational wave in 1969 [9]. This argument was rejected by the same scientific establishment that evolved into the LIGO Collaboration.

Later we will discuss how Weber was suggested to have mistaken geomagnetic disturbance for gravitational wave. So here we have the Collaboration's own assessment that coincidence may not be a strong test of gravitational wave.

Before we leave this subject I introduce a tentative point that was first raised in public by W. W. Engelhardt (see Chapter 7). LIGO's arms are each 4 km long and the wavelength of the gravitational wave is a few thousand kilometers. This makes LIGO an "electrically small" antenna (an electrically small crossed-dipole antenna would be a good analogy.) For this reason the two crossed arms of LIGO cannot be seeing different signals even when the space is being compressed along one arm and expanded along the other in the manner of a gravitational wave passing by. A linear probe of a wave that is much smaller than the wavelength is largely insensitive to the wave direction vis-à-vis its own orientation with respect to that direction. Each arm will see an averaged effect which is nearly zero.

Here is a simple way to see this. Let us consider wave motion in general. The wave propagation direction is unknown. By orienting an antenna whose size is comparable to the wavelength, one can determine the direction of the wave. However, one cannot do so if this antenna is much smaller than the wavelength where it would be termed a probe. A probe can sample the power but not determine the direction of propagation or polarization with any useful resolution. In the same way, a LIGO arm can be oriented to determine the direction of the wave when its length is comparable to the wavelength. It follows that the arm cannot be seeing different signal strength depending on its orientation when it is much smaller than the wavelength. The two orthogonal arms will sample the wave and report the same signal, if they at all do so.

To summarize Chapter 4:

CONCLUSION 1: When LIGO reports a waveform at its output, there is no way to tell what the corresponding input waveform is that was incident on the instrument from the sky. The latter waveform is in fact indeterminate. This is the end of the line for LIGO as an instrument of scientific discovery. But this feature of LIGO has been kept hidden, and then taken advantage of by slipping in a custom-tailored waveform in place of the unknown incident waveform. This is a veritable scientific shell game.

CONCLUSION 2: Where there was conducted a controlled experiment with a known input waveform and a corresponding known output waveform, the crucial proof-of-concept waveform comparison was not reported publicly. Thus "inculpatory evidence" may have been suppressed.

CONCLUSION 3: Neither the above experiment nor any other experiments reported provided any support for the assertion that LIGO measures movements down to a level of 1/10,0000-th the diameter of a proton. This is a fiction on which the discovery rests. LIGO functioned as an uncalibrated instrument when the two discoveries were reportedly made.

CONCLUSION 4: LIGO has a continuum of states of actuation, from single end-mirror vibration to both end-mirrors vibration and everything in between. The observed output waveform does not tell us anything about the state of actuation. Thus the instrument remains uncharacterized and uncharacterizable in this regard as well. There is no way to tell that the state actuation corresponds to a gravitational wave (both end mirrors in coordinated vibration) other than wishful thinking.

CONCLUSION 5: LIGO most definitely did not observe gravitational wave also because the signal it saw was attenuated in traveling from Washington to Louisiana. Gravitational waves do not attenuate like this. This diminution of amplitude is greater than can be explained away by detectors' relative geometric orientation. Additionally, this orientation-related diminution has been misestimated because of lack of knowledge of the antenna pattern. This pattern also is indeterminate, making LIGO worthless yet once again as a gravitational wave observatory.

CONCLUSION 6: Conclusions 1-4 are all redundant, considering that the LIGO waveform-measurement instrumental procedure has been greatly botched. The LIGO output waveform is not "on the sky". It a highly corrupted form of whatever signal was incident on LIGO. The two signals can bear no resemblance, given how LIGO functions. In plain language, LIGO waveforms are unsalvageable.

CONCLUSION 7 (Tentative): When the two LIGO arms are much smaller than the wavelength of a gravitational wave, they cannot detect a differential length change in the orthogonal arms when the wave passes by.

There is a television advertisement about a brand of beer that is presented as so classy that pouring it into a glass from the tap with just the right head of foam is a matter of great professional pride for the bartender. Here the bar patron is sitting on the stool in the bar car of an express passenger train barreling down the tracks. As the bartender tries to pour the beer into a glass the jerking motion of the train mars the foam head. He throws away the beer and starts with a fresh new glass. Again he is thwarted. With a grave face he excuses himself and goes out of the compartment for a bit. He comes back looking relaxed and pours the perfect glass of beer. He scrapes off the excess foam, making the head level with the rim of the glass. He offers it up to the patron with a look of great pride and even a smile just breaking. What happened? Well, it turns out that he went out and disconnected his bar car from the rest of the train which went on its way! The lone dining car was now gliding smoothly along the tracks without any jerking motion.

That is what Kip Thorne did. After years of frustrating failure and with time running out, he simply disconnected his theory car from the train of scientific procedure advancing on its tracks, and poured a fine glass of discovery quaff with a fine head of foam. But this was a classier glass of beer than Stella Artois. It cost more than one billion US dollars.

Stephen Hawking, who won the Special Breakthrough Prize in 2013, said, "This discovery has huge significance: firstly, as evidence for general relativity and its predictions of black hole interactions, and secondly as the beginning of a new astronomy that will reveal the universe through a different medium. The LIGO team richly deserves the Special Breakthrough Prize."

Yuri Milner, one of the founders of the Breakthrough Prizes, said, "The creative powers of a unique genius, many great scientists, and the universe itself, have come together to make a perfect science story."

Edward Witten, the chair of the Selection Committee, commented, "This amazing achievement lets us observe for the first time some of the remarkable workings of Einstein's theory. Theoretical ideas about black holes which were close to being science fiction when I was a student are now reality.

from the statement accompanying the award of the Special Breakthrough Prize for the LIGO discovery

5. The botched instrument: Left out in the rain

We have seen in the previous chapter that LIGO is a fundamentally flawed concept in instrumentation for scientific research. It is not a whole instrument. It does not let us transform an observed signal to an incident signal and as such it is of little use. It is like a very expensive bridge to nowhere. But as if this were not worthless enough, LIGO is also plagued by other design mistakes of most rudimentary nature. What LIGO reports is not even on the sky.

In his much-vaunted historical 1972 MIT internal report that represents the inception of the idea of the LIGO instrument, Rainer Weiss identifies geomagnetic disturbances as a potential area of concern in the measurement of gravitational waves [10]:

Geomagnetic storms caused by ionospheric currents driven by the solar wind and cosmic rays create fluctuating magnetic fields at the surface of the Earth.
...
Fluctuating magnetic fields interact with the antenna mass primarily through eddy currents induced in it or, if it is constructed of insulating material, in the conducting coating around the antenna that is required to prevent charge buildup. The interaction, especially at low frequencies, can also take place through ferromagnetic impurities in nonmagnetic materials. Magnetic field gradients cause center-of-mass motions of the suspended mass. Internal motions are excited by magnetic pressures if the skin depth is smaller than the dimensions of the antenna mass.

He then performs some numerical estimations and concludes:

Although the disturbances caused by the smoothed power spectrum do not appear troublesome in comparison with the other noise sources, the occasional large magnetic pulses will necessitate placing both conducting and high-μ magnetic shields around the antenna masses.

He goes on to suggest that the then recent discovery of gravitational waves by Joseph Weber might just have been caused by geomagnetic disturbance:

It is not inconceivable that Weber's coincident events may be caused by pulses in geomagnetic storms, if his conducting shielding is inadequate. It would require a pulse of 10^{-2} G with a rise time -10^{-3} s to distort his bars by $\Delta l/l \sim 10^{-16}$.

So here we have a clear view that from the get go the LIGO designers were aware of geomagnetic disturbances being a matter of concern. We even see the early consciousness that such disturbances can be mistaken as gravitational waves.

Turning now to the LIGO instrument as it existed in September of 2015 when the discovery was made, what electromagnetic shielding did they consequently provide to the test masses (the LIGO mirrors)? We know that the mirrors and all the associated moving structures of the instrument were encased in steel tubing. The tubes were 1.2 meter in diameter made from 304L stainless steel sheet 3 mm thick.

To determine to what extent geomagnetic disturbances can penetrate through 3 mm of steel into the region inside the tube and hit the supersensitive innards of LIGO, we may look at the skin depth for steel. This depth is the scale-length for exponential attenuation of an electromagnetic wave penetrating a metal slab (say). Such a wave can travel many skin depths through the metal and still emerge on the other side of the slab – with diminished strength.

Geomagnetic disturbances occur generally at frequencies less than 1 KHz. The resistivity of 304 steel is 72 μΩ cm. The skin depths are as follows:

Frequency	Skin Depth (mm)
1 KHz	1.4
100 Hz	4.2
10 Hz	13.5
1 Hz	42.7

It is clear that there will be substantial to nearly total penetration of the electromagnetic fields into the space of LIGO's mirrors, suspension assemblies and the associated components. The most sensitive parts of the instrument are totally exposed to the geomagnetic "elements".

How do the EM fields affect LIGO? LIGO mirrors are made of dielectric materials and utilize dielectric coatings. Hence they are not directly affected by these fields. But their support structures are. To see this we must look at the entire mirror suspension assembly (Figure V-1). This assembly – designed to isolate the mirror from mechanical vibrations of the framework from which it is suspend – contains copious amounts of metal. These metal masses would experience geomagnetically induced currents – just as Rainer Weiss had figured 43 years before the discovery. Thus LIGO was not provided that crucial scientific protection which they knew had to be provided for the experiment to succeed. It is not that they provided inadequate protection; they did not provide any protection at all. They did not improve upon Joseph Weber at all.

Note that in order to wreak havoc with the observation of gravitational waves, the EM disturbance does not have to be in the same frequency region as these waves. Whatever the wave frequency is, the wave's observation will be affected by the disturbance, whatever the disturbance frequency is.

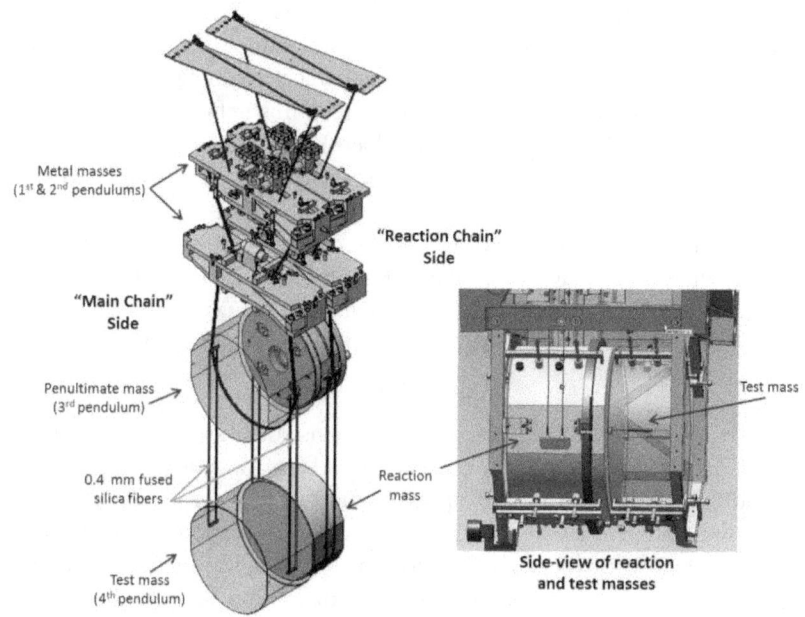

Figure V-1: The vibration isolation system for LIGO. Note the copious use of metal in this mirror suspension assembly. Use of any metal at all should have been a design no no. [*Source: LIGO*]

This type of bungling happens in large engineering projects when different people with different stakes and different responsibilities do not coordinate closely, and the coordinating project manager does not do his job. Apparently, the vibration isolation system designers went to town with their design latitude, without the physicists telling them not to use any metals. The result is that a beneficial thing becomes the bane of the project. So out there in the vast open landscape, LIGO lies totally prone to geomagnetic disturbances. It has been left out in the rain, so to speak. LIGO is in effect an observatory designed to be open to receiving geomagnetic disturbances. Gravitational waves, if they arrived, would have to be winnowed out from this cacophony of signals.

However, the importance of the electric currents induced in the metallic components by these disturbances extends beyond what Weiss had figured. Rainer Weiss forgot to take into account – in 1972 or in the 43 years since - the most important geomagnetic factor for his experiment: The static magnetic field of the Earth. The LIGO instruments sit in this magnetic field that has a strength of about 0.5 gauss. Every time a current is induced in a piece of metal, it experiences mechanical forces due to this magnetic field. So LIGO is subject to electro-mechanical vibrations. And since "vibrations" are what LIGO is trying to detect, this is fatal.

We will now demonstrate that the Earth's magnetic field is an inextricable scientific consideration for LIGO – no matter what the instrument is sensing.

LIGO designers showed no awareness of the i-cross-B (**ixB**) force in a metal carrying a current i in a static magnetic field B. We can see that this force is operative whether LIGO is being actuated by geomagnetic disturbances, mechanical effects, gravitational waves, or some combination of these.

1. *Electromagnetic Effects*: The two LIGO instruments are required to be (and reportedly are) exactly identical. Therefore the output signals at the WA and LA stations should be equally identical – warts and all - *if* they were from a gravitational wave. This was not the observation. This means that the signals *incident* on the two detectors had to be non-identical. This is true of the i_gxB force arising from an EM disturbance source high above Earth's surface, where i_g is geomagnetically induced current.

2. *Mechanical Effects*: As to actuation of LIGO by mechanical causes (seismic tremors, man-made vibrations etc.), the tubes vibrating in the field B see a current i_v (due to the vxB electric field in the metal, v being the instantaneous vibration velocity.) This in turn induces currents i_m and forces i_mxB in the suspension. So we are back in the EM regime.

3. *Gravitational Wave Effects*: But can LIGO in future detect a gravitational wave that most considerately chooses to transit the Earth when there are no geomagnetic or mechanical activities? The expansion and contraction of space (if taking place) cause metallic masses to see a time-varying magnetic field ($\partial B/\partial t$) and a consequent electric field E creating a current i_t, and the resulting force i_txB. So we're back in the EM regime.

Thus in all circumstances of LIGO the Earth's magnetic field is a parameter to be reckoned with. It was not.

Next let us look at what was said about the role of geomagnetic disturbance on LIGO at the time the discovery data traces were acquired [10]. In the following *h(t)* is the strain reported by LIGO.

Electromagnetic signals in the audio-frequency bands are also produced by human and solar sources, including solar radio flares and currents of charged particles associated with the solar wind. The strongest solar or geomagnetic events during the analysis period were studied and no effect in h(t) was observed at either detector.

This is a completely inadequate and unacceptable handling of an issue that was the most crucial pillar of their discovery. What exactly was "studied"? What sources of data/survey? What instruments? What were their properties and where/how were they deployed? What was the survey coverage – time-wise, frequency-wise, polarization-wise… etc.? A host of such minimal necessary information is missing. They are asking to you take their word for it on what will make or break the discovery. Furthermore, it is more than likely for LIGO - because of its length and orientation - to sense a geomagnetic episode that is missed by conventional antennas/detectors.

There is this other equally important point: Why look at only at the strongest events? It could just as well be a very weak event that actuated LIGO. It is probably as much a question of frequency and geometry and polarization as it is of strength when it comes to setting the LIGO suspension assembly in a mode of resonant vibration with chirp.

Eliminating geomagnetic disturbances as a cause was perhaps the most important requirement for any LIGO discovery report. Standard monitoring of such disturbances by other agencies for other purposes are not adequate. LIGO needed importantly to have its dedicated geomagnetic survey installation co-located with the gravitational wave instrument. In fact, there might have been ways to make the long LIGO tubes themselves serve as efficient antennas for sensing geomagnetic disturbances. There also needed to be sensors strategically placed inside the tubes and next to the mirrors.

What was 14 September 2015 like, geomagnetically speaking? According to the Space Weather Prediction Center (NOAA) [11]:

The geomagnetic field is expected to be at active to major storm levels on day one (12 Sep), unsettled to minor storm levels on day two (13 Sep) and quiet to active levels on day three (14 Sep).

According to other reporters, that day was a quiet to moderately active day. It is true that there was no spike recorded by anyone in geomagnetic activity that coincided with the signal observed by LIGO. But it should also be noted that long metallic structures like LIGO may pick up electromagnetic fields that more conventional sensors may miss. Here is a brief excerpt from Wikipedia [12]:

Geomagnetically induced currents (GIC), affecting the normal operation of long electrical conductor systems, are a manifestation at ground level of space weather. During space weather events, electric currents in the magnetosphere and ionosphere experience large variations, which manifest also in the Earth's magnetic field. These variations induce currents (GIC) in conductors operated on the surface of Earth. Electric transmission grids and buried pipelines are common examples of such conductor systems.

To this list can be added the new system of LIGO pipeline.

The long and short of all this is that LIGO is intimately and inextricably linked to the geomagnetic environment. Its design is most unsuitable for detecting gravitational waves. Anyone who proposes that it picked up a gravitational wave has a far higher burden of proof than the one the discoverers seemed to think they have.

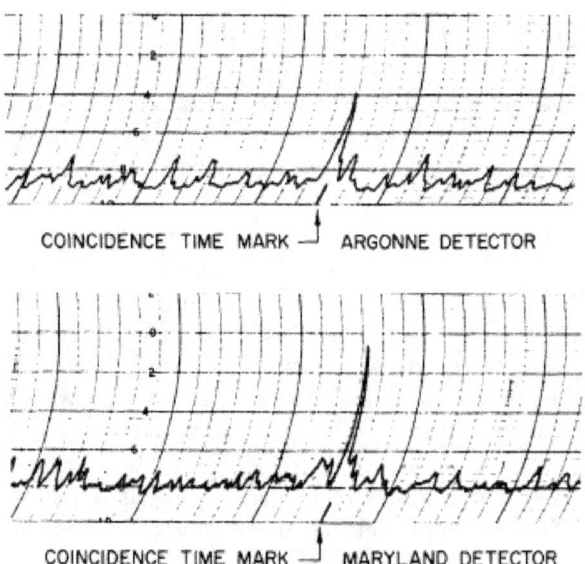

Figure V-2: The 1969 report of a possible detection of gravitational wave by Joseph Weber whose instrument was radically different from LIGO. The simultaneous detection of a signal at two stations (Argonne National Laboratory near Lemont, Illinois and the University of Maryland in College Park, Maryland) that are 1000 km apart was the justification for this discovery. However, this discovery was rejected by the physics establishment because it could not be confirmed by other experimenters.
[*Image source: APS*]

The criterion that a time coincidence of signals at two distant stations indicates a gravitational wave has been argued against by LIGO scientists themselves in the case of the 1969 gravitational-wave discovery report by Joseph Weber, as shown in Figure V-2 [9]. This establishes that coincidence is a necessary but not sufficient criterion. A geomagnetic or seismic disturbance whose source is located far enough from the detectors can register in both the detectors.

To summarize Chapter 5:

CONCLUSION 8: Rainer Weiss the LIGO instrument pioneer knew from the beginning that LIGO needed to be shielded from the geomagnetic environment. The completed LIGO instrument contains no such shielding.

CONCLUSION 9: When the LIGO discovery was reported, only lip service was paid when it came to justifying why the signal was not due to the ubiquitous geomagnetic causes to which LIGO lay fully prone.

CONCLUSION 10: Absolutely no notice was taken of the forces on LIGO test masses arising from the effect of the Earth's static magnetic field on geomagnetic currents induced in the copious metal structures inside the LIGO tubes. This is a gargantuan botch-up job by a billion-dollar, thousand-man, decades-long project.

CONCLUSION 11: Simultaneous observation of a signal at two distant stations is a necessary but not a sufficient criterion for a gravitational wave. Rainer Weiss argued that such a simultaneity reported by Joseph Weber in 1969 as a possible discovery of gravitational wave was due to geomagnetic causes. The same is true for LIGO.

Rainer Weiss has been promoted as LIGO's original architect, a kind of an Argo to Thorne's Jason. Facts on the ground prove otherwise: He is the architect of LIGO's doom. More astounding, however, is the fact that his mistakes were not spotted by some 1000 other genius-class researchers who followed him, and so rectified, over a period of 43 years. A billion-dollar supersensitive and super-high-tech instrument that they were all glorying about from their cozy offices in Caltech and MIT and NSF was left out in the rain.

Perhaps an understanding of what went on with LIGO can be gained not from science but from history. Weiss was also the Chairman of the Science Working Group for NASA's COBE Satellite. That instrument's charter was to measure the spectrum of the Cosmic Microwave Background Radiation. The satellite was launched without developing any scientific assurance on the ground about its ability to measure a spectrum of any kind at all. The successful discovery reported by COBE was the greatest science fraud in history. That project cost $400 million, spanned two decades and involved a thousand individuals.

COBE and LIGO, two unrelated cutting-edge instruments directed at unrelated grand discoveries - one a satellite in orbit and the other an observatory on ground - had one factor in common: instrument pioneer Rainer Weiss.

6. Dissidence: Slow thinking in fast times

Generally speaking, when a scientific discovery is announced, the reaction within the relevant scientific establishment ranges from applauding acceptance to polite reserve. There is nowadays another forum operative on the Internet consisting of people who are outside that establishment, and often outside the academia. They may include disaffected scientists with strong positions, science enthusiasts, and amateurs who think they have something important to say on the subject. All this is known, and is not very germane to the present discussion.

But there was an important development in this regard with respect to LIGO. There was a complete polarization in the views of the two forums. The first one expressed their unanimous acceptance and offered their ultimate praise. There was not a hint of doubt. The second one – starting on the very day the discovery was announced – showed great skepticism. Some of their comments were astute and applicable. It is this polarization that I find of interest, hinting at a far larger issue related to the plight of science.

Therefore I will mention some of the comments from the second forum, the dissidents in this case. I will then present a few representative comments from the establishment side.

Claes Johnson is Professor of Applied Mathematics at the Royal Institute of Technology in Stockholm, Sweden. On 12 February 2016, the day after the discovery was announced, he made a blog post part of which was as follows [13]:

Absurdity of Modern Physics: LIGO Gravitational Wave Detection as Ill-posed Problem

We understand that an expansion-contraction of the Earth of the size of an atom nucleus diameter as an effect of a "ripple in the fabric of space-time" was detected during 0.25 seconds, and from this observation the conclusion is drawn by computer simulations and modelling that this extremely minute effect as a "ripple in the fabric of space-time", was the result of a very specific extremely gigantic invisible explosion 1.3 billion light years away shining brighter than all stars in all galaxies for 0.25 seconds in the form of gravitational waves.

We see a combination of a biggest possible cause/input and a smallest possible effect/output in a certain mathematical model. The conclusion comes from using this mathematical model in inverse form, where a smallest possible signal is used to identify a biggest possible origin of the signal.

This means that the mathematical model in inverse form is extremely ill-posed and as such cannot be used to draw conclusions. To do so requires that all alternative explanations of the zero signal can be eliminated, and it is then not enough to just say that no other explanation immediately suggest themselves, that is to draw conclusions from ignorance with the precision of the conclusions increasing as the ignorance or stupidity grows.

I like this comment because of its fundamental logical clarity. Basically it says this: How can you probe infinity with zero? This is a serious consideration at the root of the LIGO approach. LIGO's difficulty or credit is not measuring a small quantity. It is value of this small quantity that is at issue.

And then we have Otto E. Rössler, Professor of Chemistry at the University of Tübingen in Germany. His post on 18 February 2016 reads partly as follows [14]:

Does a "Fraudulent Joke" stand behind the Discovery of Gravitational Waves?

Second, the data are interpreted as a length change in the 4 kilometers long detector arms, functionally enlarged by a factor of more than ten through back and forth reflections. The measured length change then is less than one ten-thousandth of the diameter of a proton — more than 24 orders of magnitude below the 4 kilometers of the instrument. This unprecedented sensitivity means also that no physical effect has ever been excluded to a higher accuracy! If the observed minute length change was real, gravity waves would be a million times weaker than originally anticipated by Joseph Weber whom I once met. Never has a tinier result been presented as a discovery.

Again, this is a perceptive observation. Note especially the point about the need for excluding all other explanations down to a fantastic level of smallness of their roles. Indeed, one might feel in his gut that almost any conceivable effect anyone can think up will produce a signal larger than gravitational waves.

Commenting on this post on 4 May 2016 in the same blog site, Wolfgang W. Engelhardt, a retired professor of Max-Planck-Institut für Plasmaphysik, noted:

The LIGO-publication must be classified as "fraud", since it claims a strain of 10^{-21} between the two arms caused by a gravitational wave, but it does not document the same strain under controlled conditions caused by radiation pressure on the mirrors. This was one of the methods described as being applied for the calibration of the system. The calibration curve, however, is withheld from the public!

There is another intrinsic problem with this particular "gravitational wave". It had a typical frequency of 150 Hz belonging to a wave length of 2000 km, if it travels at c. As the distance between beam splitter and reflecting mirror was 4 km, there was practically always the same phase on these objects, i.e. there was no reason to move them in different directions against each other. In this case there is no relative shift and a phase difference between the two arms of the interferometer cannot occur.

Maybe I am wrong, but I cannot see, how a relative displacement between beam splitter and mirror can occur, unless the wavelength is comparable with the arm length.

These are most insightful comments. Later Engelhardt would offer further critical comments on LIGO in an "Open Letter to the Nobel Committee for Physics 2016" [15]. Here is the abstract:

The Nobel Committee is informed that according to Professor Karsten Danzmann (Albert Einstein Institut) the LIGO detectors are not calibrated as expected from the statement in the discovery paper: "The detector output is calibrated in strain by measuring its response to test mass motion induced by photon pressure from a modulated calibration laser beam [63]". The claim that gravitational waves have been detected is not substantiated experimentally, since direct calibration data, namely mirror displacement as a function of laser power moving the mirrors, are not published.

On 28 March 2016 Demetris T. Christopoulos of the Faculty of Economics, National and Kapodistrian University of Athens, posted a paper on the Internet. Here is the abstract [16]:

A detailed critical review of reported event GW150914 that LIGO/VIRGO collaboration announced as gravitational waves and black holes observation

ABSTRACT

On 11 February 2016 the analysis of advanced LIGO event GW150914 was published and claimed to be the first direct observation for gravitational waves and black holes at the same time. At this more detailed review we read most of supporting papers and make remarks on both experimental and theoretical basis. First we show that time given from GPS is being treated as an absolute entity. Then we indicate why such a detection is a contradiction to the Principle of Equivalence. After we focus on the Numerical Relativity section, where we find that field equations cannot be solved numerically, unless they have been converted to either elliptic or hyperbolic form. That is achieved by using a very strict set of initial assumptions about the topology of 'spacetime' and by dividing it in 3+1 components. Software used for supporting main waveform templates is a black-box one without any information given about its basic characteristics, thus all presented numerical data cannot be checked. Scientific ethics in project LIGO is strongly violated by using the so called blind injection software and hardware procedures.

As a result of the excess use of blind injections it is necessary an international independent scientific Committee to be established in order to check in depth the whole LIGO/VIRGO project and PRL as its official publisher journal. Question about what exactly happened on September 14, 2015 remains open.

This author would later perform follow up studies that would bear out his comments above [17].

On 19 February 2016 blogger Mark Mahin made extensive comments on the discovery. Here are portions of his comments [18]:

There are two parts of the LIGO announcement. The first is the claim that gravitational waves were detected (which as we have seen is on rather shaky ground). The second is the claim that these waves were caused by a merger of distant black holes. The second claim is speculative, and is not well supported by the evidence.

The scientists had no direct evidence that the claimed signal came from a distant black hole. What they basically did is to do some calculations showing a hypothetical scenario by which a black hole merger might have produced the described signal. But that is not at all the same as showing that such a hypothetical scenario was the actual cause. Given a gravitational wave observation, there are always many possible ways of explaining it astronomically. We are reminded here of the BICEP2 affair, when scientists triumphantly claimed that the signals they detected came from the dawn of time. It was later shown that just such a signal could have been produced by dust. There was no way of even telling from which direction the LIGO signal was coming, so the scientists did nothing to show that the signal came from an exact spot in the sky where black holes are known to exist.

You would think that facts such as these would cause our scientists to be cautious. Science is supposed to require repeated observations, not one-shot wonders. But the scientific community has thrown caution to the wind in this matter. Based on a single rather questionable observational event, the scientific community has acted as if LIGO is proof of gravitational waves and proof of a black hole merger. The first claim is shaky, and the second claim is extremely shaky.

These are thoughts that should have occurred to every thinking scientific observer, but clearly they did not occur to the establishment scientists in jubilation around LIGO.

On or about 16 March 2016 blogger Shannon Sims made comments portions of which are reproduced below [19]:

Lack of third-party environmental monitoring

The published paper mentioned that each observatory site is equipped with an array of sensors to monitor "environmental disturbances and their influence on the detectors". Nowhere is there any indication that any other sensors outside of the detector sites were consulted, even those operated by any of the many institutions that are part of the LIGO collaboration.

Gravitational wave angle limitations

The LIGO detectors were built to detect gravitational waves that travel perpendicular to them, stretching and compressing the lengths of their long L-shaped arms. However the time difference between the two signals was less than 7 milliseconds.

This means the waves would have had to travel more parallel to the detectors which would have made them significantly more difficult to detect.

These are examples of questions that arise once one starts to delve into the discovery logically. That did not happen with the establishment.

But why is it that different people find different things wrong with the same discovery report? The answer is simple: When an apple is rotten to the core, you can find fault with its taste, its smell, its looks, its texture, its feel … etc. The LIGO discovery is likewise flawed to the core.

Now let us look at a few representative comments by establishment's prominent members in order to compare and contrast this view with the aforementioned view.

A science popularizer seemingly with Yoda-like wisdom, Michio Kaku is a professor at the City College of New York. He waxed eloquent on the future prospects of the discovery (implied to be a done deal) in the Wall Street Journal of 12 February 2016 [20]:

Now we are witnessing the third great revolution in telescopes, the use of gravity waves to open a new chapter in astronomy. For the first time, waves from the very instant of creation might be observed, giving us "baby pictures" of the universe as it was born. High-school textbooks may have to be rewritten to incorporate the new discoveries coming from this third generation of telescopes.

In an article in The New York Times, 11 February 2016, Lawrence Krauss, a professor at Arizona State University and another science popularizer, bought the LIGO discovery in toto and proceeded also to look forward from there [21]:

What more can we learn about the universe from a stupefying experimental feat observing a stupefying wonder of nature? The answer is anyone's guess. Gravitational-wave observatories of the future will be able to explore the exotic features of black holes. This may shed light on the evolution of galaxies, stars and gravity. Eventually, we may be able to observe gravitational waves from the Big Bang, which will push the limits of our current understanding of physics.

In June of 2016 Brian Greene, a professor of physics at Columbia University and a noted science popularizer, said in an interview conducted by the Kavli foundation [22]:

For a while now, it has gotten to a point where scientists would speak as if the existence of black holes was a done deal, but that was always a little quick.
The evidence now has taken a major leap forward, because there are these beautiful computer simulations of the gravitational wave shapes that would be produced if two black holes were orbiting each other.
To have that signal actually wash by the LIGO detectors in Louisiana and Washington, to have a wave state that so closely matches the result of the supercomputer calculation of Einstein's theory, that really cinches the case. Now you've got this direct agreement between the numerical simulations of orbiting massive black holes and the observations of gravitational waves.

It is important to note that the Kaku-Krauss-Greene set is not a collection of run-of-the-mill science reporters. The former are practicing scientists of stature and are seen as official explainers for the physics establishment. The world accepts their narrative in implicit trust. The media promotes them as such.

But even as a myriad issues about LIGO became apparent to the outside-establishment thinkers from the get go, this set was busy installing and entrenching LIGO before the citizens of the world from the get go. They were doing so with great stylistic flair.

I have to say this: The comments of the dissidents are characterized by a natural inquisitiveness that is the very soul of science. The views of the Kaku-Krauss-Greene set are those of town criers. It is a great irony of our times that the latter group today posits itself as the inspiration for the world's young, and no one takes any note of the former group.

And this brings us back to the issue of the polarization of views among the LIGO scientific establishment and the non-establishment dissidents. If the former were right, there would be nothing noteworthy about this polarization. We would say: So what else is new? But the case is the exact opposite. Just as some elderly people are easy mark for boiler room scammers, the prevailing intellectual elite was easy mark for the LIGO scammers. But luckily, the dissidents caught on to them quickly.

There is only one way to rationalize this. It is to heed the lesson of history enunciated by Arnold J. Toynbee: *The cause of the fall of a civilization occurred when the cultural elite became a parasitic elite.*

7. Assignment of blame: *Enlightened people...*

We begin at the beginning, with Kip Thorne. He is one of the world's foremost expert on the theory of gravitation. It is natural that he should have developed a keen interest in observing gravitational waves and further exploring them. His zeal in raising funds for this and building an enormous scientific empire is admirable. His persistence and insistence in continuing the funding when times were tough are not necessarily self-serving. But after the observation of 14 September 2015 came to his knowledge, it was incumbent on him to bring the same level of intellectual ardor to bear on the scientific process that unfolded. The resemblance of the observed data to what his theoreticians and numerical modelers had custom-designed was uncanny. Any scientist would feel uncomfortable with such perfection.

Kip Thorne is not an experimentalist in any sense of the term. But even within his own expertise he could have reached the conclusion that something was clearly amiss. Right in front of his eyes were stark contradictions. The instrument stood as one of unprecedented high quality to have measured the wave – complete with inspiral, chirp and ringdown. But the same instrument also said that at the same time that the gravitational wave, after traveling 1.3 billion light years, had become substantially attenuated in making the journey from Washington to Louisiana. He could also see that there was no scientific way to demonstrate that his theory waveform was what had passed through the instrument to become the observed waveform. There was a total disconnect in the "chain of custody" for the waveform right at the instrument input face.

Still and all, if he had rolled out a *possible* discovery, then no harm no foul. That is not what he did. He forcefully pushed this as the discovery of black hole merger with the loudest loudhailer available. Later when the prizes were being showered on him for his discovery, he accepted them as a self-assured discoverer. And right now, because of him, and entirely false branch of science is entering the school textbooks around the world.

Kip Thorne of Caltech should be assigned the primary accountability for the LIGO scam and all that is associated with it and all that flows from it.

Rainer Weiss the instrument pioneer is multiply at fault. But for now let us only cite the fact that he left LIGO wide open to receiving all kinds of geomagnetic disturbances. He is the one who identified early on that the instrument needed protection in this regard. But right in front of his eyes the LIGO design was led astray in this regard. So what else is new? Right in front of his eyes, the COBE Satellite scientific instrumentation was led astray. He was the Chairman of the COBE Science Working Group. His responsibility was proximate and immediate – then as now.

Rainer Weiss of MIT should be assigned the primary accountability for the LIGO instrumental botch up.

The National Science Foundation has all along been the financial backer of LIGO. I have seen the expression "unwavering" backer. There is nothing wrong with this. What is important to recognize here is that the funding mechanism is controlled not by NSF staffers and leaders, but by the scientific establishment. This control is exercised through the National Science Board and a host of other review panels. So, sorting out the historical fault of NSF in the LIGO matter would be an entire project unto itself. This is outside the scope here.

What we can do here is to look at the conduct of the present-day NSF, especially starting with the announcement of the discovery. That event was hosted by NSF's current director France A. Córdova who is an astrophysicist herself. She did everything in her sphere of influence to make the event a worldwide public-relations success, including prompting the issuance of tweeted congratulations by US President Barack Obama and by the White House Office of Science and Technology Policy. The latter issued a congratulatory statement which contained a parenthetical message seemingly to tell any scientific critics not to bother [23]:

(The months between the observation in September and yesterday's announcement were spent in intensive examination and re-examination of the signal by the team, and by peer reviewers, in order to ensure beyond reasonable doubt that there could be no explanation other than gravitational waves for what had been observed.)

This report gloried over the long-haul role of NSF in patiently nurturing LIGO to fruition. The NSF Director also arranged for a kumbaya-style laudatory Senate Hearing on LIGO. Later she would preside over the highly visible event of international significance, the signing the Memorandum of Understanding between the US and India on LIGO on 31 March 2016. After that, following a meeting between President Barack Obama and Prime Minister Narendra Modi in June 2016, a statement was issued by the White House which cited LIGO as a model for Indo-US collaboration in science and technology [24]:

Expanding Cooperation: Science & Technology and Health

35) The leaders affirmed their nations' mutual support in exploring the most fundamental principles of science as embodied in the arrangement reached to cooperate on building a Laser Interferometer Gravitational Wave Observatory (LIGO) in India in the near future and welcomed the formation of the India-U.S. Joint Oversight Group to facilitate agency coordination of funding and oversight of the project.

All this and more helped highlight before the world the solid backing and affirmation of the discovery from the US Government and the US Senate, and the Indian Government and the Indian Union Cabinet. A few feeble dissident voices that arose on the Internet immediately following the discovery report soon fell silent. After all, who can stop the march of such naked power?

The NSF director has been a willing or unwitting party to a great perversion of science. This damage to science may be irreversible. Of course no blame attaches to her for the historical developments. But leading India down the path that would install a most expensive but totally bogus technology showpiece on Indian soil was not right. Even if the LIGO leaders were egging her on to do this, a worthy chief executive in her place would take note of, and try to get to the bottom of the Internet dissidence on LIGO. Córdova is in that office not only to be a cheerleader for the establishment. She is there to be a prudent custodian of the nation's scientific aspirations. Internet dissidence is very much a part of what her office, of all places, needs to take into account. As I have noted in the Falsifiers book, I believe that Michael Griffin, former NASA chief executive, did exactly that with regard to the COBE Satellite. He understood leadership.

We need to touch on the so-called science popularizers, the likes of Brian Greene, Michio Kaku and Lawrence Krauss. Unlike science reporters who are paid to faithfully report what the establishment says, the science popularizers are each their own man, and are themselves active scientists. They have the ability to sense a scam and understand it especially when someone raises or flags an issue. It is their place then to at least introduce appropriate caveats into their narrative. Yet, these individuals went all out to push the discovery as a done deal, in prominent venues and with crusaders' zeal.

The science popularizers are accountable for spreading and promulgating bogus science around the world, whether unknowingly through lack of competence or knowingly with deliberate purpose. Do we really need them?

But perhaps the most bizarre of all the corruptors of science have been the prize givers. No sooner was the discovery announced than Kip Thorne was created a TIME100, a shorthand for TIME Magazine's one of the most influential 100 people in the world for 2016. This seemed to have been done in an awful hurry for a process that has an annual cycle. Then came the $3 million Special Breakthrough Prize, special because it was given out of turn. Then came the Gruber Cosmology Prize, followed by the Shaw Prize in Astronomy and the Kavli Prize in Astrophysics. A couple of these prizes seemed to have waived their own nomination deadline to make the award to LIGO. Kip Thorne and Rainer Weiss each would receive more than a million dollars, going to their personal bank accounts.

While all this was going on there was copious talk in the media of a LIGO Nobel Prize in 2017. The 2016 nomination deadline had passed when the discovery was announced. The 2017 prize was said to be a sure thing, the only unpredictability being who all would share the prize.

2016 PRIZES AWARDED FOR LIGO DISCOVERY
Awarding panels

SPECIAL BREAKTHROUGH PRIZE

Nima Arkani-Hamed	Alexei Kitaev	Adam Riess*
Lyn Evans	Maxim Kontsevich	John H. Schwarz
Michael B. Green	Andrei Linde	Nathan Seiberg
Alan Guth	Arthur McDonald*	Ashoke Sen
Stephen Hawking	Juan Maldacena	Yifang Wang
Joseph Incandela	Saul Perlmutter*	Edward Witten
Takaaki Kajita*	Alexander Polyakov	

GRUBER COSMOLOGY PRIZE

Robert Kennicutt	Frans Pretorius	Andrew Fabian
Sadanori Okamura	Helge Kragh	Owen Gingerich
Rashid Sunyaev	Subir Sarkar	Martin Rees

SHAW PRIZE IN ASTRONOMY | KAVLI PRIZE IN ASTROPHYSICS

Jiansheng Chen	Richard McCray	Paolo Caselli
Peter Goldreich	Brian P. Schmidt*	Carlos Frenk
Michel Mayor	Mats Carlsson	Fiona Harrison
	Claude Canizares	

*Nobel Laureate in physics

Table VII-1: Some physicists who staked their reputation on the LIGO discovery.

The moving force behind these prizes was the physics establishment. Table VII-1 shows the people involved in awarding these prizes. This list is a veritable Who's Who in Physics today. Now, is it believable that these physicists had no inkling that anything was the matter with LIGO? Why is it that people who are largely hardcore theoreticians and mathematical physicists are making an assessment of a futuristic engineering project in an uncharted territory of scientific experimentation? These backers of the said prizes helped entrench LIGO further, if further entrenching was needed.

There is no reason to think that the deceased philanthropists Peter Gruber, Fred Kavli and Run Run Shaw had any special agenda for science in mind. We must look at those who are administering the endowments, and their advisors. These people are accountable for a gross perversion of science. The trustees of these endowments are not doing their duty.

The long-traditioned physics journal *The Physical Review Letters* published by American Physical Society has provided the scientific imprimatur on the discovery. Because of its accumulated historical record of publishing great discoveries, this journal has come to be considered solidly credible.

The first LIGO discovery the journal published [3] was a very complex multi-disciplinary matter with aspects of advanced engineering that spilled beyond the boundaries of traditional physics. A truly thorough review of this paper might be an interminable process when it had to be published promptly. The paper came from a large group of premier scientists from premier institutions, operating a mega-science project. The result reported was of the highest physics significance. So, all things considered, the editors may not have had an option to not publish this paper.

So the journal cannot be faulted for publishing the first paper. But as I discussed, the second paper [4] was bizarre. Publishing it as a second discovery demonstrated bad judgment on the part of the editors. But to place this in context, we should remember how these same editors acted with respect to the BICEP2 discovery. There again, the publication of the first report [25] of that bogus discovery was probably justified for the reasons stated above. But the second paper [26] they published was a whitewash job. It was supposed to report that the BICEP2 instrument was a botch up. Instead, it reported that BICEP2 was a solid instrument which happened to observe the wrong thing, solidly. This was a clear, concerted and deliberate cover up. That act was repeated with LIGO.

Following the publication of the LIGO discovery the journal received other submissions that are essentially run-of-the-mill idea papers that take off from LIGO. There was no need to publish these papers in a journal dedicated to rapid publication of timely and important papers as well as discovery reports. And yet the journal published such papers [27-30]. This is an all-too-transparent strategy on the part of the editors to bolster the LIGO mystique and maintain its scientific visibility.

The quality of leadership at *The Physical Review Letters* has fallen far. It now conforms to the quality of the scamming scientific groups they serve – groups that have become endemic in this day and age.

Finally, we will touch on the conduct of two individuals I have covered in the Falsifiers book: George Smoot (2006 Nobel Prize for physics for the COBE Satellite map of the early universe) and Brian Schmidt (2011 Nobel Prize for physics for discovery that the expansion of the universe is accelerating.) Both discoveries were bogus. But now they both rose to endorse the new bogus discovery with their Nobel Laureate stature.

Immediately following the report of the discovery, Brian Schmidt was in the Australian media plugging the discovery in superlative terms. This was effective plugging, given that Australia had a big role in LIGO. He was also instrumental in awarding the Kavli Prize to LIGO. George Smoot made an entire presentation at the 2016 Lindau Nobel Laureate Meeting on the LIGO discovery, averring that the discovery was a done deal. So what we have here is past scam discoverers rising to authenticate latter-day scam discoverers. A certain culture perpetuates itself.

This is a good point to remind ourselves of George Orwell's observation: *Enlightened people seldom or never possess a sense of responsibility.*

"Only the federal government can invest in this kind of long-range, high-risk research," said Dr. Maria Zuber, Board Member and vice president for research at MIT. "In this case the National Science Foundation, with a successful outcome far from assured, nurtured an idea with small grants, followed by major investments over decades. The magnificent discoveries reported here are only the tip of the iceberg of what will be learned as this new observatory takes its place alongside NSF's electromagnetic telescopes and the IceCube Neutrino Observatory at the South Pole."

"I'm overwhelmed by the implications of this discovery," said Dr. Anneila Sargent, professor of astronomy at the California Institute of Technology and chair of NSB's Committee on Programs and Plans. "It is clearly only the first of many breakthroughs we'll hear from the LIGO team. It also serves to highlight the value of partnerships between NSF and universities; the long-term partnership between this federal science agency and MIT and Caltech accomplished what was once thought to be unachievable."

from the National Science Board statement on LIGO discovery

8. LIGO India: A nation disserved

The visionary leaders of LIGO long had it in their minds to set up an outpost as far removed distance-wise from the two US stations as possible. This would offer a long baseline for better identifying the source direction and also provide an independent distal corroboration for any US observations. It later emerged that behind this logical plan they had a far greater evangelical plan in mind: To set up a worldwide network of gravitational wave astronomy observatories and usher in a new era of scientific exploration that would immortalize the pioneers.

The original thought was to set up this new station in Australia. However, in 2011 - after years of discussions - the Australians declined, citing budgetary constraints. But Australian scientists remained invested in the American LIGO project. India was also involved in these discussions and now rose to the occasion. The negotiations on an Indian LIGO began to take place. But the process was moving at a snail's pace.

A LIGO India (also called Indigo, for Indian Initiative in Gravitational-wave Observations) consortium was constituted. The roster of members was developed logically enough, including scientific experts, institutional representatives, and so on. Among the Indian LIGO scientists who rose to prominence in the media were Bala Iyer of the Tata Institute of Fundamental Research; Sanjeev Dhurandhar and Tarun Souradeep of the Inter-University Centre for Astronomy and Astrophysics (IUCAA); and Somak Raychaudhury, director of IUCAA which in time became the main LIGO institute in India. Also, the renowned cosmologist Jayant Narlikar at the IUCAA emerged as an avuncular inspiration to the rest of the team.

Some three dozen Indian scientists in India also began to contribute to LIGO research. The contribution was largely in theory, computation and signal analysis. There does not seem to have been any significant involvement of these scientists in LIGO observation or instrumentation.

Two American scientists of Indian origin were advising India on hosting LIGO. They are Abhay Ashtekar, a professor of physics at the Pennsylvania State University engaged in theoretical research; and Rana Adhikari, a professor at Caltech and a prominent LIGO scientist who has some instrumentation background. He was a doctoral student of Rainer Weiss. None of these two scientists was in a position to certify that LIGO was a properly functioning instrument from the viewpoint of their own expertise.

Following the 11 February 2016 report of detection of gravitational wave by LIGO, the snail's pace suddenly accelerated to that of a racehorse. Not even a week elapsed before the Indian Government and the Union Cabinet approved LIGO in principle on 17 February 2016. The fast-paced movement continued until a Memorandum of Understanding was signed between the United States and India on 31 March 2016 in Washington DC.

The next step was to zero in on an appropriate site for Indian LIGO. The scientists had already narrowed the list down to just two sites, one in the state of Rajasthan and the other in Maharashtra.

The strangest thing for me is that I could not find anywhere any indication that Indian engineers had conducted an independent study of the LIGO instrumentation. The Indian LIGO later hired an engineering firm to survey the prospective sites as to their suitability for LIGO. This shows that they have the ability and the awareness to retain outside consultants as and when needed. But when this was crucially needed to assess LIGO engineering before advancing with the proposal, they did not think to do so. Given the success of India in space-faring technology, one would have thought that such expertise would be readily available indigenously.

Think about that: A huge engineering installation using cutting edge technology and beyond, and advancing into uncharted territory of scientific experimentation was to be commissioned in India. It was to be an enormous open-ended national commitment. What was the first and foremost task for the Indian LIGO movers? It was to conduct the nation's own assessment of the science and engineering involved. If such a study were conducted, it would have opened a can of worms.

Instead, theoreticians and computation types were leading the show on the strength of their own expertise. The engineering part was proceeding on implicit trust that the Americans had to have done things right.

It is to be noted here that for scientific reasons the Indian LIGO instrument had to be an exact copy of the American version. There was no room for radical modifications. So India's choice was to accept LIGO as is or reject it. If an engineering study gave rise to serious issues, they would have to reject it.

In the event that an engineering study *was* conducted in India, it clearly did not produce the right result.

After the discovery was announced, this fact in and of itself was seen as certifying the LIGO instrument's quality and functionality. A bogus instrument paired with a bogus discovery was passed off as a fine instrument successfully producing a fine discovery. The Indian LIGO scientists made that representation to their Government and their Union Cabinet that LIGO had been spectacularly vindicated. They got approval with spectacular haste.

This representation was scientifically unwarranted. I need only cite Chapter VI on dissidence, showing that many observers outside the LIGO scientific establishment had instantly spotted multiple major faults with the discovery report.

The Indian LIGO scientists reportedly had a big role in the second LIGO discovery. They forcefully announced to the media that this second discovery had removed all doubt.

The Indian LIGO scientists, by creating all this extra hoopla around the LIGO discoveries, have given this scam an extra boost in the international arena. They in effect knowingly or unknowingly aided and abetted in installing the scam discovery as legitimate science. The LIGO scam has fleeced the American and some international taxpayers to the tune of $1.1 billion thus far. Now comes the turn for the Indian taxpayers to be fleeced, thanks to their own scientists.

But it is also true that I may be proved wrong as to LIGO being a bogus instrument. The Indian LIGO will succeed if they devoutly follow the American script. Great discoveries will follow. That script is to systematically erect a highly ramified body of false science which facilitates the making of mythical heroes. Cosmology is the most convenient arena for this purpose. Individuals or small groups do not have any ability to recheck measurements made by a billion-dollar observatory or a satellite. Nor will the arguments offered by these people – solid but not presented in the format of a conventional paper in a scientific journal – will be considered 'properly submitted'. So the enterprise is quite safe.

Whether America is exporting this culture of scamming or the Indians are importing it entirely of their own volition, only the insiders can say. It is probably a little of both.

Jayant Vishnu Narlikar – a fine Indian scientist (a theoretical cosmologist) who deserves a fine legacy – has emerged to be a spiritual leader of Indian LIGO. He has also been an ardent lifelong critic of Big Bang Cosmology, a field I exposed to have been constructed on faulty instrumentation and scammed reports of discovery therefrom. When I was an undergraduate student in Calcutta, he suddenly burst upon the Indian scene as a great discoverer (in regard to his work with Fred Hoyle on supernova creation field and gravitation field.) I remember making an effort to go the venue where he gave a captivating lecture. He was young, handsome and charismatic. He made a great impression on my young mind just as I was considering taking up astronomy and astrophysics. Later, around about 1974 when he visited La Jolla, he and his family graciously accepted the invitation to come to our cramped campus quarters for dinner. We felt honored and graced. I certainly hope that his role in Indian LIGO will not tarnish his main legacy.

9. Concluding remarks

Shortly after the LIGO discovery was announced and I examined it, I suggested on my blog site that the LIGO scientists retract the discovery and mount the Captain Asoh Defense [31]. This is still the best way out.

Over the past several years I have investigated and reported on a number of grand scams in physics, perpetrated mostly in the murky frontier land of cosmology. But in doing so I have sensed, slowly and flittingly, the shaping of a longer-term trend. That trend is towards a denaturing of science by entrenching in it false elements through raw establishment power and highly clever machinations. This is a historical decline of the scientific civilization, unfolding right before our eyes. We are watching our doom engulf us in real time. But we are doing nothing. It is as though we are in a trance. We are lying languidly in squalid opium dens and a watching phantasmagoric razzmatazz playing out before us.

A thousand men and women – represented to us as the finest talent ever gathered under one umbrella, represented by the august scientific establishment, by the great institutions of learning, by the raucous media, and indeed by the geniuses themselves – had designed some intellectual version of a *son et lumière* extravaganza. They were performing their acts on a 360-degree open-air stage under a full moon in dazzling costumes, and we were all watching spell-bound. Occasionally a child exclaimed: Why are they all naked?! We took no notice. We were in an opiate stupor.

In venues like Lindau Nobel Laureate Meeting and World Science Festival, this scam science is nicely packaged as inspirational science and genius science, and delivered to the young generation. They buy into this.

Like the Roman Empire, physicists rose to great heights. Like the Roman Empire, physicists are now going through a phase of decline. The causes of their declivity: arrogance, hubris, puffery and limitless thirst for public adulation. These negative qualities are being aided by a decline in the positive qualities: breadth and depth of scientific imagination (as distinct from mathematical calisthenics), intellectual honesty (as distinct from bureaucratic and procedural cover of correctness), and a sense of service to the society (as distinct from a love of intellectual auto-stimulation.)

It is difficult – if not impossible – for even the science-conscious, thinking citizen to figure out that this decline is happening. The reason for this is that the three avenues through which this citizen could perceive the decline are actually corrupted.

THE QUALITY METRICS

The public perceives the quality of the scientific establishment largely from the high profile honors and accolades that they give themselves.

These metrics have declined along with the establishment. A scientifically aware citizen may have sometimes thought that physics Nobel Prizes were being given for small matters, and then he may have dismissed that thought, saying 'Nah! What do *I* know?!' But he was actually right. Even bogus and fraudulent discoveries have begun to be anointed now. Today subtle but intense campaigns – political-style – are waged for the physics Nobel Prize in behalf of "candidates".

THE MOUTHPIECE

The public's main source of information is the media, and especially, the so-called science reporters. These are actually the touts for the establishment. Whatever line they are being fed, they will run with it. They will report this, get paid, go to Safeway and put food on their table. There is no way that these people are going to be the first to tell you that the quality is declining.

THE ENABLERS

Today, almost all of the research in physics is funded by the Government. It has set up its own huge bureaucracy to interact with the scientific establishment and administer the funds. This is actually a cozy gravy boat for both parties, and no one is going to rock it. The same is true with the legislative body. They communicate only with the establishment, and ignore any input from outside the establishment. If somehow an outsider can get them to look at a complaint against the establishment, they will consult a member of the establishment.

The members of the physics establishment at large, meaning the vast numbers in day-to-day drudgery outside of the privileged and powerful groups, know and understand all this. But they do nothing. They simply watch, like so much listless cattle upon a wasteland.

For there is really nothing to do. If some people somehow are able to expose a grand scam, the same naked power that promoted and protected it will now move in to whitewash over it. The scammers will face no accountability of any kind. They will be saved harmless. They will simply go mum, as will the media that thus far promoted the scam. Taxpayers will be out billions of dollars. But who cares? They are after all the most inconsequential players of all the players.

Let us conclude with an apt summary of LIGO which can done by paraphrasing a famous soliloquy:

It is a tool built by an idiot, full of bells and whistles, signifying nothing.

NOTES AND REFERENCES

[1] A concise history and description of LIGO is available in the Wikipedia article titled LIGO.

[2] De, B. (2015): The Falsifiers of the Universe, *Createspace Independent publishing Platform*, Charleston, SC.

[3] Abbott, B. P. and coauthors (2016): Observation of Gravitational Waves from a Binary Black Hole Merger, *Phys. Rev. Lett.* **116**, 061102.

[4] Abbott, B. P. and coauthors (2016): GW151226: Observation of Gravitational Waves from a 22-Solar-Mass Binary Black Hole Coalescence, *Phys. Rev. Lett.* **116**, 241103.

[5] De, B. R. and Nelson, M. A. (1992): Ultrabroadband Electromagnetic Well Logging: A Potential Future Technology. *Transactions of the SPWLA Thirty-third Annual Logging Symposium,* Paper A.

[6] Abbott B. P. and coauthors (2016): Calibration of the Advanced LIGO detectors for the discovery of the binary black-hole merger GW150914, *arXiv:1602.03845 [gr-qc]*.

[7] LIGO Scientific Collaboration, Advanced LIGO, 2014: https://arxiv.org/pdf/1411.4547, LIGO-P1400177-v5

[8] Weinstein, A. J. (xxxx): Overview of LIGO Core Technologies, LIGO-G000165-00-R.
Retrieved at:
http://elmer.caltech.edu/ph237/week12/weinstein.pdf

[9] Weber, J. (1969): Evidence for Discovery of Gravitational Radiation, *Phys. Rev. Lett.* **22**, 1320.

[10] Muehlner, D. J. and Weiss, R. (1972): Gravitation Research, *Research Laboratory of Electronics (RLE) at the Massachusetts Institute of Technology (MIT)*, Date Issued: 1972-04-15.
Retrieved at:
https://dspace.mit.edu/handle/1721.1/56271

[11] The LIGO Scientific Collaboration and the Virgo Collaboration (2016): Characterization of transient noise in Advanced LIGO relevant to gravitational wave signal GW150914, *arXiv:1602.03844 [gr-qc]*.

[12] Refer to Wikipedia article titled *Geomagnetically induced current*.

[13] Johnson, C. (2016): Absurdity of Modern Physics: LIGO Gravitational Wave Detection as Ill-posed Problem, Claes Johnson on Mathematics and Science, *claesjohnson.blogspot.com*
Retrieved at:
http://claesjohnson.blogspot.com/2016/02/absurdity-of-modern-physics-ligo.html

[14] Rössler, O. (2016): Does a Fraudulent Joke stand behind the discovery of Gravitational waves? *www.researchgate.net*

Retrieved at:
https://www.researchgate.net/post/Does_a_Fraudulent_Joke_stand_behind_the_Discovery_of_Gravitational_Waves

[15] Engelhardt, W. W. (2016): Open Letter to the Nobel Committee for Physics 2016, *www.researchgate.net*
Retrieved at:
https://www.researchgate.net/publication/304581873_Open_Letter_to_the_Nobel_Committee_for_Physics_2016

[16] Christopoulos, D. T. (2016): A first critical review of event GW150914 observed by LIGO detectors Livingston and Hanford, *www.researchgate.net*
Retrieved at:
https://www.researchgate.net/publication/294705090_A_first_critical_review_of_event_GW150914_observed_by_LIGO_detectors_Livingston_and_Hanford

[17] Christopoulos, D. T. (2016): A detailed critical review of reported event GW150914 that LIGO/VIRGO collaboration announced as gravitational waves and black holes observation, *www.researchgate.net*
Retrieved at:
https://www.researchgate.net/publication/299454440_A_detailed_critical_review_of_reported_event_GW150914_that_LIGOVIRGO_collaboration_announced_as_gravitational_waves_and_black_holes_observation

[18] Mahin, M. (2016 February 19): LIGO Doubts Will Persist Unless Replication Occurs, *Future and Cosmos*.
Retrieved at:
http://futureandcosmos.blogspot.com/2016/02/ligo-doubts-will-persist-unless.html

[19] Sims, S. (2016 March 16): PROBLEMS WITH THE LIGO GRAVITATIONAL WAVE DISCOVERY, *PLASMA PICS*.
Retrieved at:
http://plasma.pics/problems-with-the-ligo-gravitational-wave-discovery/

[20] Kaku, M. (2016 February 12). Riding Gravity Waves to the Big Bang and Beyond. *Wall Street Journal.*
Retrieved at:
http://www.wsj.com/articles/riding-gravity-waves-to-the-big-bang-and-beyond-1455318961

[21] Krauss, L. (2016, February 11). Finding Beauty in the Darkness. *The New York Times.*
Retrieved at:
http://www.nytimes.com/2016/02/14/opinion/sunday/finding-beauty-in-the-darkness.html

[22] Greene, B. (2016 May 18). Gravitational Waves: An interview with Brian Greene (interview). *www.kavliprize.org*
Retrieved at:
http://www.kavliprize.org/events-and-features/gravitational-waves-interview-brian-greene

[23] Holdren, J. P. (2016 February 12): Statement on the Detection of Gravitational Waves, *www.whitehouse.gov*
Retrieved at:
https://www.whitehouse.gov/blog/2016/02/12/statement-detection-gravitational-waves

[24] The White House (2016 June 7): JOINT STATEMENT: The United States and India: Enduring Global Partners in the 21st Century, *www.whitehouse.gov*
Retrieved at:
https://www.whitehouse.gov/the-press-office/2016/06/07/joint-statement-united-states-and-india-enduring-global-partners-21st

[25] Ade, P. A. R. and coauthors (2014): Detection of B-Mode Polarization at Degree Angular Scales by BICEP2, *Phys. Rev. Lett.* **112**, 241101.

[26] Ade, P. A. R. and coauthors (2015): Joint Analysis of BICEP2/Keck Array and Planck Data, *Phys. Rev. Lett.* **114**, 101301.

[27] Cardoso, V. and coauthors (2016): Is the Gravitational-Wave Ringdown a Probe of the Event Horizon? *Phys. Rev. Lett.* **116**, 171101.

[28] Bird, S. and coauthors (2016): Did LIGO Detect Dark Matter? *Phys. Rev. Lett.* **116**, 201301.

[29] Gerosa, D. and Moore, C. J. (2016): Black Hole Kicks as New Gravitational Wave Observables, *Phys. Rev. Lett.* **117**, 011101.

[30] Sasaki, M. and coauthors (2016): Primordial black hole scenario for the gravitational-wave event GW150914, *Phys. Rev. Lett.* (Accepted for publication 29 June 2016).

[31] Here is the Captain Asoh Defense:

THE CAPTAIN ASOH DEFENSE

Japan Airlines Flight 2 was a flight piloted by Captain Kohei Asoh on November 22, 1968. The plane was a new Douglas DC-8 named "Shiga", flying from Tokyo International Airport to San Francisco International Airport. Due to heavy fog and other factors, Asoh mistakenly landed the plane in the shallow waters of San Francisco Bay, two and a half miles short of the runway. None of the 96 passengers or 11 crew were injured in the landing.

The passengers all evacuated the plane on lifeboats. The plane came to rest on solid ground 10 feet below the water, leaving the forward exits above the waterline. It was not severely damaged and was recovered 55 hours after the incident, transported to the airport on a barge. United Airlines refurbished the aircraft for service at their maintenance base at the airport, at a cost of roughly $4 million USD. The aircraft was returned to JAL on March 31, 1969, where it was renamed "Hidaka" and continued in service to JAL until 1983.

Asoh, when asked by the NTSB about the landing, reportedly replied, "As you Americans say, I fucked up." In his 1988 book The Abilene Paradox, author Jerry B. Harvey termed this frank acceptance of blame the "Asoh defense", and the story and term have been taken up by a number of other management theorists.

Asoh was demoted to First Officer, went through further ground schooling, and continued to fly for JAL until his retirement.

Selected portions from Wikipedia article Japan Airlines Flight 2

ABOUT THE AUTHOR

Bibhas De is the author of the 2015 book *The Falsifiers of the Universe*, subtitled *Big Bang Cosmology: The first fraud in the final frontier*. It is a stark investigative report on how Big Bang Cosmology was installed through a series of experimental or observational studies that were each some combination of sham, scam, and fraud.

Bibhas De received his Bachelor's degree with Honors in Physics from Presidency College, University of Calcutta; Master's degree in Astronomy from the University of Michigan, Ann Arbor, with training in radio astronomy which he received under the guidance of a pioneer in the field, Fred T. Haddock; and Ph. D. degree in Applied Physics from the University of California at San Diego in 1973 as a doctoral student of physics Nobel Laureate Hannes Alfvén. De has since worked in both academic research and industrial R&D. In the former area he has many scientific publications. In the latter area he holds a number of patents. His experimental background ranges from small-scale laboratory experiments to large-scale field experiments. His engineering background includes Antenna & Microwave Theory and Techniques applied to satellite communication and other applications.

www.ingramcontent.com/pod-product-compliance
Lightning Source LLC
Chambersburg PA
CBHW070333190526

45169CB00005B/1870